Biopolitical Governance

Global Political Economies of Gender and Sexuality

Series Editors: Nicki Smith, Adrienne Roberts and Juanita Elias

This series brings together scholarship from leading and emerging scholars working on the intersections between gender and/or sexuality in political economy. It seeks to move beyond the 'blindness' of International Political Economy to feminist, gender, trans*, queer and masculinity studies in order to more fully capture the complex and contested transformations associated with globalization, capitalism and neo-liberalism.

Titles in the Series

Biopolitical Governance: Race, Gender and Economy, edited by Hannah Richter
Realising Justice for Sex Workers: An Agenda for Change, edited by Sharron A. FitzGerald and Kathryn McGarry
Livable Intersections: Re/Framing Sex Work at the Frontline, by Sara M. Kallock
Heavy Metal, Politics and Feminism: Sexy or Sexist, by Heather Savigny

Biopolitical Governance

Race, Gender and Economy

Edited by
Hannah Richter

ROWMAN &
LITTLEFIELD
INTERNATIONAL
London • New York

Published by Rowman & Littlefield International Ltd
Unit A, Whitacre Mews, 26-34 Stannary Street, London SE11 4AB
www.rowmaninternational.com

Rowman & Littlefield International Ltd. is an affiliate of Rowman & Littlefield
4501 Forbes Boulevard, Suite 200, Lanham, Maryland 20706, USA
With additional offices in Boulder, New York, Toronto (Canada), and Plymouth (UK)
www.rowman.com

British Library Cataloguing in Publication Data
A catalogue record for this book is available from the British Library

ISBN: HB 978-1-7866-0271-8
PB 978-1-7866-0270-1

Library of Congress Cataloging-in-Publication Data

Names: Richter, Hannah, 1990- editor.
Title: Biopolitical governance : race, gender and economy / edited by Hannah Richter.
Description: London ; New York : Rowman & Littlefield International, Ltd, [2018] | Series: Global political economies of gender and sexuality | Includes bibliographical references and index.
Identifiers: LCCN 2018005167 (print) | LCCN 2018019289 (ebook) | ISBN 9781786602725 (electronic) | ISBN 9781786602718 (cloth : alk. paper) | ISBN 9781786602701 (pbk. : alk. paper)
Subjects: LCSH: Biopolitics. | Sex—Political aspects. | Race—Political aspects. | Racism—Political aspects. | Emigration and immigration—Political aspects. | Refugees—Social conditions. | Immigrants—Social conditions.
Classification: LCC JA80 (ebook) | LCC JA80 .B546 2018 (print) | DDC 320.01—dc23
LC record available at https://lccn.loc.gov/2018005167

Contents

Introduction

The Two Bodies of Biopolitics

Hannah Richter

> Malina still drinks his coffee. 'Oi' can be heard from across the other rear window. I have stepped towards the wall, I step towards the wall, hold my breath. I was supposed to write a note: it wasn't Malina. But the wall opens, I am in the wall, and for Malina only the crack can be visible, which we have seen for a long time. He will think that I left the room.
>
> —Ingeborg Bachmann (2008, 296)

Ingeborg Bachmann's *Malina* ends with the nameless female first-person narrator withdrawing to a crack in the wall and disappearing. Bachmann leaves no ambiguity regarding the fact that the life of the disappearing woman, in its physical and linguistic expression outside the wall, was moulded by the patriarchal status quo of power relations crystallised into her relational connection to two men: 'I have lived in Ivan and I die in Malina' (2008, 296). The violence exercised through these relations is what ultimately causes her death: '[i]t was murder' (ibid., 298). Filled with references to psychoanalysis and Simone de Beauvoir's female other, Bachmann's work is usually read as a literary reworking of the social and sexual victimisation and discursive silencing of women in the political context of post–World War II Austria. Bachmann herself intimately connects the othering and coercion operating along the embodied line of gender which runs through her work to the segregating thanatopolitics that had just surfaced under the Nazi regime. The victims of the Shoah and Bachmann's female characters share the experience of being 'entirely a body on which the history/and not one's own, is being played out' (Bachmann 2000, 60).

Yet there is another, more active dimension to the final act of disappearance through the crack in the wall narrated by Bachmann, closer aligned with what she referred to as her 'politics of weakness' (ibid., 152) which is aimed at rupturing the 'pre-established harmonies' (ibid., 37) of power relations. The withdrawal to the crack happens in the absence of Malina, the male alter ego of the protagonist ('He will think that I left the room'). Robbed of a female body as its receiving end, the patriarchal power of Malina is forced to disappear together with the voice of the female narrator: '[s]teps, always Malina's steps, more muted the steps, most muted steps. A standstill' (Bachmann 2008, 298). Bachmann's crack, understood in this sense, operates parallel to the surface crack-up which Gilles Deleuze detects as running through the works of Émile Zola, F. Scott Fitzgerald and Malcolm Lowry in *The Logic of Sense*. The crack-up releases a dramatising openness in which 'the death Instinct is reflected inside an open space, perhaps even against itself' (Deleuze 2004, 370), rupturing the relations of onto-epistemological reproduction which keep in place the status quo of common sense and good sense.

However, while Deleuze's crack-up emerges and draws meaning from the psychoanalytic depth which lies beneath the surface of sense, Bachmann's crack is the appearance of a both reproductive and resistant materiality with the governed body (of the female narrator) as its operative centre. More precisely, there are two bodies which make up the governmental system in *Malina*, or rather, there is one body split into two. Both Malina, the sociopolitically immanent patriarchal power, and the resistant force of the female narrator reside in the body gendered, sexually normalised and violated through the exercise of the former, the same power which had enacted the deadly racialisation of the Holocaust. But because this bodily materialisation plays a constitutive role for the governmental force personified in Malina, it must perish together with the body it governs, the body of the governed.

The idea that governance takes place as the relational moulding of two bodies, continuously split and continuously interconnected, lies at the heart of Foucault's theory of biopolitics and the Foucaultian political and social thought it continues to inspire (Lobo-Guerrero 2016; Kelly 2015; Lemm and Vatter 2014; Vatter 2014). Biopolitics, as conceptualised by Foucault (2008; 1978; 1976), is the governance of a clearly delineated, homogenised, totalised population which is made live beyond the institutions of sovereign governance and the Fordist economy. It is only in so far as politics gives life to and sustains this singular population-body as the One of biopolitics that it, in turn, is endowed with functional legitimacy for the measuring, modulating and monitoring exercise of its life-giving power.

But because this conceptualisation reveals biopolitics as contingent upon the logic that a governmental administration of rights and freedoms is beneficial for each rationally calculating citizen, this population governance is at the same time inherently tied to the liberal subject as the individual bearer of these rights. For Foucault, biopolitics and liberalism—both in its theoretical underpinnings and its political-economic exercise—are intrinsically linked. 'How can the phenomena of "population", with its specific effects and problems, be taken into account in a system concerned about respect for legal subjects and individual free enterprise? In the name of what and according to what rules can it be managed?' (Foucault 2008, 317). Foucault's biopolitics must thus be understood as a technique of government which takes place in and reproduces itself through the smooth interrelation of two bodies which constitute the biopolitically governed: the population as the unit to which its specific governmental techniques are geared and which they target, and the body of the individual subject as the unit where these productive governmental technologies take effect and become visible, constituting the aggregate of the population as a multiplicity of, in turn, productively aligned subjective wills.

The play at Ernst Kantorowicz's *The King's Two Bodies* (1997) evoked by the idea of the two bodies of the governed is of course not accidental. Foucault (1991) himself uses Kantorowicz in *Discipline and Punish* to explore the same constitutive function of bodily materiality for disciplinary power which Bachmann's *Malina* alludes to. But in a more extensive way, Giorgio Agamben (2012) draws on Kantorowicz in *The Kingdom and the Glory* to unpack the apparatus of biopolitics in its self-reproducing capacity which resides in the dramatic split of governmental power into its own two bodies, one constituent, immaterial and immortal and one mortal, material and constitutive. Emergent modern political power, whose legitimacy is in a state of permanent precariousness, requires a primary, immanent body which functions as the secularised unmoved mover of the force exercised upon its subjects. In the absence of an ontological foundation, this immanent power requires the second body to for its constitutive materialisation—be it through the display of glory on the body of the king or the exercise of force on the bodies of the subjects. By visibly receiving transcendent governmental force, the body thus grounds the immanent power from which it emerges in the first place in a continuous process of *ex post facto* constitution (Agamben 2012, 121–56).

This constant splitting and mutually reproductive interrelation of the two bodies of governmental power is, for Agamben, the functionality and purpose of the theologically rooted political economy. Similar to Bachmann, Agamben (2016) assumes that it cannot be ruptured through active resistance, which would only feed into its dialectic operationality always already materialised in the body of the governed constituted through its biopolitical split from the political realm. As in Bachmann, it can only be escaped through a seemingly passive withdrawal of the constitutive biopolitical subject to an inside, in-between space of standstill where the relations of biopolitical reproduction are nullified to allow for the emergence of resistant creativity.

Both Foucault and Agamben thus theorise biopolitics as a relational network connecting two bodies, the capital 'B' Body of the immanent whole and the small 'b' body of its constitutive parts. While held in motion by the immediate materiality of the latter, biopolitics can only become effective and sustain itself as the economic apparatus which interrelates both bodies in a mutually reproductive way. Zooming in on this economic functionality of biopolitics, the One of biopolitics can be unpacked as a third body—the body of capital relations. Following Althusser's (2016, 65–66) reading of Marx, the body of capital 'in which all relations coexist simultaneously and support one another' (Marx 1963, 110) is society in its epistemological conditionedness by and sustainment of capitalism as a self-reproducing relational economy. Capital is here not an essence or capacity, but a relation which produces further capital by connectively extending its totalised relational Body. This constant self-extension is absolutely necessary to sustain not only the epistemic integrity of the former through the repeated delineation of its bodily realm but also its economic vitality against the background of constantly falling profit rates.

The Body of capital internalises new matter, life and labour force. The combination of its economic and epistemic relations ensures that individual lives are only sustained through labour processes in so far as they also nurture and sustain the abstract, relational Body of capital they are connected to. It is in this sense that Marx (1976, 342; 367) famously describes capital as vampiric. Itself without stable figure or essence, in constant need for sustainment through external connections, vampiric capital makes the labour force, which is relationally included in its totality, live. But it does so only in so far as the Body of capital at the same time extracts this life for its own reproduction, turning bodily labour or bodily matter into capital. What Marx sets up through the idea of the labourer as victim of the vampire of capital is thus yet another version of the dialectic of the two bodies of biopolitics. This time, the dialectic is centred on the dynamic movement of bodily production itself: the relational individuation of the governed body of economic value-creation and the totalising feeding of the vampiric Body of value extraction—the Body of capital. To Foucault and Agamben's theorisation of the politically constitutive interrelation of the two bodies sustained by the materiality of the governed body, Marx's perspective adds that the productive employment of this materiality takes place through an economic apparatus which, from the side of the biopolitical One, individuates the bodies of the governed as exploitable and consumable and aligns them within a movement of introversive recharging directed towards the totality of the former.

Despite targeting different dimensions within the operational logic of governance, all three iterations of the two bodies of biopolitics unpack a dialectical mechanism of self-reproduction which makes the Body of the biopolitical One—framed either as governed population, governmental power or the socioeconomic relations which sustain the former—live through its productive materialisation in a multiplicity of governed bodies. But importantly, as in Bachmann, the functioning of the biopolitical apparatus is conditional upon the immediate, visible materiality of the body of the governed. Malina only lives because the female first-person narrator supplies him with a body to exercise his power upon. In Bachmann, the body of the governed is a gendered, sexed, heterosexual body—ethnically ambiguous but aligned with the racialised bodies of Holocaust victims through the governmentally productive violence which is exercised upon it. It is a body already formed as biopolitically employable through the governmental line of sex which, in the words of Foucault, lies 'at the pivot of the two axes along which developed the entire political technology of life. . . . Sex was a means of access both to the life of the body and the life of the species. It was employed as a standard for the disciplines and as a basis for regulations' (Foucault 1978, 145–46).

In Foucault, the politically productive body of the governed is moulded along two central biopolitical lines which align its productivity with the life and health of the population Body—the line of sex and the line of race. Focusing on the two hinges of the Foucaultian biopolitical apparatus, this volume explores the body of the governed as the nodal point of biopolitical governmentality. The chapters collected here retrace the intricate and complex lines of biopolitical production carving out a governed body which holds in motion the dialectic apparatus constitutive of the biopolitical One. Unravelling the broad notions of 'sex'[1] and 'race' in their productive relation to not just one but all of the biopolitical totalities differentiated above, the lines of embodied biopower are on the one hand drawn from the work of Foucault and the Italian and French theorists indebted to and aligned with it. On the other hand, the chapters of this volume analyse these lines of biopolitical moulding as they dominate the events, decisions and discourses of past and present politics: gender, sex, sexuality, ethnicity, citizenship status and economic value. Of course, this collection does by no means provide a complete map of embodied governmental production. Left out are, amongst others, the biopolitical lines of (dis)ability, bodily 'fitness' or mental health importantly highlighted by recent scholarship (Mitchell and Snyder 2016; 2015; Tremain 2015; Apostolidou and Sturm 2016; Larsen 2011; Doran 2017; Tarizzo 2011).

However, in line with this research, the volume aims to emphasise the persistent role of material, visible embodiment for a biopolitics which is increasingly researched as the technological, particularly computational surveillance, coded provision and algorithmic alignment of aggregate information up to the infamous 'big data' (Rouvroy 2016; 2011; Lecat-Deschamps 2015). As the underside of what Antoinette Rouvroy (2016; 2011) labels 'algorithmic governmentality', the chapters collected here show how every population-level governance must ultimately take effect in the production of the embodied life of its subjects—and relationally draws its power and political legitimacy precisely from the capacity to do so. Whether it is in the moulding of neuronal networks, the racialising monitoring of airport passengers or the experiences of LGBTQ citizens in the United States after the election of the Trump administration facilitated by the algorithmic spread of fabricated or highly perspectivist news stories, the governmental power executed here only comes into being as such when and because it directly affects the embodied life of the citizens. Their bodies are the ready-made material surfaces to actualise a constitutive force which lacks all ontological ground. In his 'Postscript on the Societies of Control', Gilles Deleuze uses an excerpt from a dystopian short story written by Félix Guattari to illustrate this relationship between algorithmic modulation and embodied political effect.

'Félix Guattari has imagined a city where one would be able to leave one's apartment, one's street, one's neighborhood, thanks to one's (dividual) electronic card that raises a given barrier; but the card could just as easily be rejected on a given day' (Deleuze 1992, 7)—affecting the specific bodily movement of the controlled citizen, which is limited or inhibited. What follows for the analysis of biopolitics, in the words of Maurizio Lazzarato, is that '[i]f power, in keeping with this description, is constituted from below, then we need an ascending analysis of the constitution of power dispositifs, one that begins with infinitesimal mechanisms' (2002, 103). Unpacking these infinitesimal mechanisms as the specific lines which interconnect the two bodies of biopolitics in their different dimensions is the aim of this volume. But how can the biopolitical mechanisms of racialisation and gender/sexual normalisation be understood in their specific operational logic? Importantly, the chapters collected in this volume are united in their scepticism towards an epistemologically or politically fixed, clearly discernible, one-dimensional and necessarily divergent functionality of the biopolitical lines of race and gender. They thereby work against and complicate simplifying tendencies towards a disconnected, bipolar understanding of biopolitics which can be traced back to Foucault's work.

In his *Society Must Be Defended* lectures, Foucault (2008, 52–62) genealogically unpacks how the idea of race came to be used as an instrument of violent political struggle—first between geographically distinct but equally powerful opponents, then by a hierarchically superior state force to reproduce its status. Imposed upon history to replace the antique, holistic logic of sovereignty, Foucault shows how the modern assumption that a racially underpinned antagonism constitutes the essence of politics itself, famously held by Carl Schmitt, transforms from a tool deployed by the sovereign at will to a technology of decentred, dispersed governmentality. It delineates the population to make live from its inferior outside, which can be left to or made die, in such a way that this process of exclusion is beneficial for the life of the former.

> What in fact is racism? It is primarily a way of introducing a break into the domain of life that is under power's control: the break between what must live and what must die. . . . It is a way of separating out the groups that exist within a population. It is, in short, a way of establishing a biological-type caesura within a population that appears to be a biological domain. This will allow power to treat that population as a mixture of races, or to be more accurate, to treat the species, to subdivide the species it controls, into the subspecies known, precisely, as races. That is the first function of racism: to fragment, to create caesuras within the biological continuum addressed by biopower. Racism also has a second function. Its role is, if you like, to allow the establishment of a positive relation of this type: 'The more you kill, the more deaths you will cause' or 'The very fact that you let more die will allow you to live more'. (Foucault 2008, 254–55)

The biopolitical line of race thus allows neoliberal governance to limit the realm of individual rational choosers whose lives can only be controlled in so far as they are attended to and—supposedly—improved. It provides governmentality with a constitutive outside against which the biopolitical One of the governed population Body is constantly produced anew. On the contrary, the biopolitical line of gendering and sexual normalisation which Foucault carves out in the first volume of *The History of Sexuality* operates through a logic fundamentally different from but complementary to the segregating line of race. When, in the nineteenth century, power relations break away from the political sovereign and his disciplinary institutions as their reproductive centre, the functionality of governance correspondingly shifts from the distribution of death to the productive administration of life. The reproductive health of the individual body, aggregated to constitute the Body of the bourgeois elite and then the population as a whole, becomes the epistemic ground and target of governmental power (Foucault 1978, 140–44).

This double biologisation of the two bodies of the governed sets in motion a biopolitical apparatus that 'has to qualify, measure, appraise, and hierarchize, rather than display itself in its murderous splendour' (ibid., 144). What is interesting to note here is that Foucault equates this biopower to make live with the immanent, symbolic power that outlives its material dramatisation in the body of the king in Kantorowicz. 'We must at the same time conceive of sex without the law, and power without the king' (ibid., 91). The biopolitical technique of gendering and sexual normalisation allows governmental power to operate and sustain itself in a concealed, indirect way, conditioning the reproductive health of individual bodies rather than manifesting itself in their visible, dramatic coercion. It does not end or replace governmental violence but situates it within a self-reproductive 'ontologico-political apparatus' (Agamben 2016, 133). This apparatus ensures that governmental violence is grounded and legitimised by the improvement of life within the governed population, and always already aligned with the reproduction of the former. The biopolitical line of sex, in Foucault, is what makes the dialectic of the two bodies work.

The dialectic of the two bodies of biopolitics is thus held in motion by another mutually reproductive binary set up between race and sex/gender as specific lines and logics of biopolitical production. While the exclusive, segregating logic of race is constitutive of the population to be governed, it is exercised by a constituent power which is grounded in the ability to nurture and sustain the Body of this population as a whole. This collection does not aim to question the accuracy of Foucault's genealogical development of race and sex as two distinct, complementary lines of biopolitical governance. It certainly also does not seek to undermine or take away from the theoretically illuminating and politically relevant critical research which is being undertaken through Foucault's theoretical lens with regard to each line, the gendered and sexually normalised production of neoliberal, economically exploitable subjectivity (Repo 2015; Cheah 2013; Inhorn 2009) as well as the—often violent—social, political and economic exclusions performed and legitimised by the racialising underpinnings of past and present governmental action (Smith and Vasudevan 2017; Welch 2016; Pooja and Chow 2013).

However, the chapters collected here firstly tie in with recent attempts (Balfour et al. 2016; Weheliye 2014; Davis 2012) to understand how the different lines of biopolitical governance interrelate, intersect, ground each other or create tensions as they materialise on the bodily surface of the governed subject. But secondly, and even more importantly, they undo and complicate the binary of thanatopolitical race and reproductive gender/sex as two separate, distinct and fixed logics of biopolitical governance by unpacking theoretical ambiguities as well as historical and geographical contingencies and dynamics in the production of the body of the governed. As pointedly summarised by Jemima Repo in her contribution in this volume, 'the functions of life and death are not exclusively tied to the apparatuses of sexuality and race, respectively, but rather these apparatuses are deployed in different historical moments and contexts to perform the administration of the life of the species, often involving radical reversals of discourse, strategy and tactics' (2018, 42). The collection explores how racialisation, gendering and sexual normalisation set up a dynamic which constitutes a multidimensional body of the governed as the reproductive centre of a biopolitical totality which is itself multidimensional: governed population, immanent governmental power and aggregate of capital relations.

The chapters in the first section of the volume, *The Politics of Life beyond Foucault*, identify new trajectories and points of development within the works of Foucault. They productively add to the framework of Foucaultian thought to analyse biopolitically produced and biopolitically productive bodies in innovative, multidimensional ways. In the first chapter, Marco Piasentier dissects Foucault's theory to show that the status of bodily life in Foucault, and the biopolitical thought which draws on his work, is ambiguous and unresolved. Without committing to either side, Piasentier argues, Foucault's work is located in an intermediary space between a sociohistorical position which attributes ontological and political primacy to the productive force of culture on the one hand, and a biological-vitalist prioritisation of the embodied force of nature or life on the other hand. Linking Foucault's ambiguous position to Davide Tarizzo's conceptualisation of life as error as a possible resolution, Piasentier points to the difficulties and pitfalls which come with any attempt to overcome the Foucaultian ambiguity while not losing sight of the both produced and productive quality of any assumed primacy within the ontologico-political apparatus of biopolitics.

Ultimately, he suggests that a biopolitical theory which seeks to critically unpack rather than enforce the operationality of the apparatus of biopolitics must situate both itself and its objects of analysis within the mutually productive relations between culture and nature held in motion by the former. In her chapter, Jemima Repo also re-examines Foucault's writings to argue against the assumed marginalisation of race and racism in his biopolitical theory detected by Ann Laura Stoler (1995) and others. However, Repo notices a certain theoretical ossification in the way biopolitical racialisation and sexual normalisation are understood. As a consequence, what is missed out is how a dynamic, spatio-temporally conditioned biopolitics can draw on sex and race interchangeably and flexibly to both delineate and advance the life of the governed population. Employing Deleuze and Guattari's concept of de-/reterritorialisation, Repo shows how the 'life function' through which biopolitics operates adapts to changing historical conditions so that the biopolitical lines of race and sex can function normalising within the governed population as well as thanatopolitical at its margins.

Hidefumi Nishiyama then analyses the influence of geographic situatedness on the functioning of biopolitics. Going beyond the comparatively sparse remarks on the biopolitical role of colonial governance in Foucault's own work, Nishiyama unpacks the racialising techniques, such as fingerprint indexing, which were deployed by the Japanese colonial government in China in the nineteenth and twentieth centuries. On the one hand, his analysis reveals parallels to Western colonialism rooted in the link between the advancement of scientific knowledge and the governmental techniques available to the colonial state. But on the other hand, Nishiyama highlights how the cultural-geographic context of Japanese colonialism conditions the emergence of a racialising governance which is, in many ways, distinct from the rule of Western colonial empires—for instance, in its focus on hierarchical normalisation within the governed population rather than thanatopolitical exclusion from the former.

While these chapters show how different biopolitical lines, nature and culture, race and sex, space and time, can work parallel and enforce each other in the production of governed bodies, intersectionality studies insist that they can also conflict, create tension and thereby point towards directions for resistance. In his analysis of 'Sex-Classification Policies as Transgender Discrimination', Heath Fogg Davis (2012) shows how gendering, economic classification and racialisation enforce the governmental scrutiny of transgender citizens facilitated by each line. But he also argues that the intersectional collision of governmental force, in its conflicting patterns of inclusion and exclusion, can spark political change. Reflecting on the experience of a black transgender woman who legally challenged the state-mandated sex identification on a Philadelphia bus, Davis notes that

> [a]s a black woman she was 'in place' as a bus rider, given that the majority of bus riders in Philadelphia are black women. But as a black transgender woman, she was frequently, and unpredictably, deemed 'out of place' in this moving public place. . . . Her racial appearance and class position were edited out of her formal legal complaint, but they cannot be redacted from the story of how she came to be at the center of a gender identity discrimination case in the city of Philadelphia at the dawn of the twenty-first century. (2012, 55)

In a biopolitical spin to Foucault's (1997, 167) famous assertion that resistance comes first—is always already present in the relations through which governance is exercised and reproduced—the chapters collected in this volume explore how both governmental and resistant power are produced at the bodily intersection of race, sex, gender, economic value and citizenship status. Importantly, the possibility for resistance is here located in the dynamic biopolitical lines which situate the body of the governed in a mutually constitutive relationship with governmental power, not in the materiality or life of this body itself. Despite its focus on the open-ended, productive materiality of the body of the governed, this collection only cautiously ties in with the recent interest in an ontologically primary, creative force of matter or life that can be summarised under the label of new materialisms.

It recognises that the governed body is always and thoroughly moulded by the ontologico-political apparatus of biopower which can only be dialectically reproduced through any conceptualisation of a governmentally untainted outside of life. The materiality of the governed body cannot be separated from its political constituted- and constitutiveness. For this reason, the chapters collected here resist any simplistic vitalist suggestion that there is something to be found within the organic materiality of life or the right-bearing subjectivity it constitutes which lies outside the apparatus of biopolitical governance and can thus serve as a secure, readily accessible ground to challenge and overcome its workings. On the contrary, this collection suggests that it is precisely because the body of the governed forms the centre of the biopolitical apparatus that a small shift in the position of the body within its relational grid, micropolitical in the sense of Deleuze, can alter the course of biopolitical production and open up pathways for theoretical challenge and practical-political resistance.

Recent developments within biopolitical thought seem to converge on the governed body as a space and source of resistance to the dialectic reproduction of the biopolitical One in exactly this sense. In his late works, Foucault (1988) himself suggests an 'aesthetics of existence', the intertwined, creative experimentation in both thought and embodied practice, as a way to transform the thoroughly biopolitical care of the self into a tool for political resistance. Giorgio Agamben continues this Foucaultian pathway in *The Use of Bodies* (2016). Parallel to the disarraying withdrawal of the narrator's female body to the crack in the wall in Bachmann's *Malina*, Agamben suggests withdrawal to an abstract 'form-of-life', which is embodied, but not sociopolitically productive, as a way to undo the ontologico-political split constitutive of biopower. Roberto Esposito's work on immunity which explores thanatopolitical governance through the conflation of its legal and biological underpinnings takes a slightly different direction. Yet again, in *Immunitas*, Esposito (2011) draws on bodily examples of mutually affirmative immunisation taking place on the microlevel of cellular interaction to work out a type of immunitarian relationality which is alternative to the thanatopolitics of the state operating on the political macrolevel, and which can be constitutive of a different mode of communal life. While the sketched-out theoretical pathways of Foucault, Agamben and Esposito have certainly been incorporated into biopolitical scholarship, explorations of how these can be used to understand and alter concrete political practice remain few, and still mainly abstract-theoretical (Prozorov 2017; Milchman and Rosenberg 2011; Deutscher 2010).

Responding to this lack of practical-political and economic engagement and analysis, the chapters collected in the second and third sections of this volume provide insights to how the self-reproductive apparatus of embodied biopolitics is situated and becomes visible in the relations of global capitalism as well as the discourses, actions and processes which shape contemporary politics. But importantly, they highlight how these biopolitical mechanisms can always at the same time be challenged through the same, intersecting lines of embodied productivity. In the second section of the volume, *Mapping Intersectional Geographies of the Body: Race, Gender, Sexuality, Economy,* both my own and Greg Bird's chapters topically explore the racialising contours of migration politics. The alignment of the biopolitical scrutiny and segregation of migrants with political racism follows Etienne Balibar's identification of the migrant's body as the focus point of a 'new racism . . . of the era of "decolonization"' where it is embodied culture which functions 'like a nature, and it can in particular function as a way of locking individuals and groups a priori into a genealogy, into a determination that is immutable and intangible in origin' (1991, 22–23).

In a theoretical exploration of governmental responses to the highly politicised European Union 'refugee crisis' of 2015 and 2016, my chapter is aimed at illustrating the politically constitutive function of politically and legally excluded, spatially confined refugees as *homines sacri* through Agamben's work. While illustrating the politically constitutive function of this exception, in a double sense, I not only challenge a certain political naïveté in Agamben's own theorisation of the refugee movement, but also the static, spatial underpinnings of his theory of constitutive power and resistance more generally. Against this background, I propose an alternative, more flexible theoretical framework which draws on Deleuze's cinematic analysis of movement-images to unpack how transgressive refugee movement can function politically reproductive for a governmental power which does not persist on the grounds of spatial stasis but through the constant alignment and modulation of flux.

Unravelling the racialising as well as gendering functionality of value-extraction from migrant bodies in the bioeconomy, Bird shows how Canadian 'guest worker' programmes politically feed the Body of capital while at the same time immunising it from the need to politically and legally include guest workers in the governed population. Doing so, he draws out the bioeconomic underpinnings and implications of Esposito's theorisation of sovereign states as immunised communities founded on notions of (the) proper(ty). The life force of the migrant other—extracted as cheap, both racially contoured and gendered labour, but excluded from the relations of the community it sustains—is what makes the Body of the Canadian society live. But at the same time, if uncontained, it constitutes a threatening vitality against which the former must politically immunise itself through a system of laws and executive patterns. They target the governed body of the migrant so that its labour force can be extracted, but sickness or reproductive relations don't feed back into and threaten to alter the socioeconomic Body of the population.

While the body of the governed is here the body of the migrant made politically—and economically—productive through racialisation, Maria Fannin analyses the bioeconomic use of the compartmentalised female body through the emergent tissue economy. Fannin's chapter opens up a new dimension to unpack the body of the biopolitically governed in its conditioned productivity, pointing out how mechanisms of economically individuating gendering can operate on the microlevel of blood and tissue samples. Fannin shows how gendering notions of nurturing and caregiving not only underpin and facilitate the vampiric extraction of value from compartmentalised governed bodies, but at the same time have the effect that the body parts individuated as sources of capital are more likely to belong to female donors. Yet, through the same gendering line, the bioeconomically exploited female bodies—for example, if pregnant—are produced as inferior in economic value for and within the Body of capital and its population.

Charlie Yi Zhang then unpacks how racialising, gendering and sexually normalising lines of biopolitical governance—in their geographic specificity already pointed out by Hidefumi Nishiyama—interrelate intersectionally to produce an embodied subjectivity which reproduces the ideals of Western neoliberalism as 'aspiring fantasy'. Providing an auto-ethnography of his own experience of growing up in China during the period of rapid economic liberalisation in the 1980s and 1990s, Zhang explores how a subject constrained through embodied lines of classing, gendering and sexually normalising governance which enforce each other turns paradoxically to the neoliberal ideas of individual excellence and self-realisation as pathways for liberation. As Zhang shows, the actualisation of this freedom, intrinsic and instrumental to the ontologico-political operationality of biopolitics, is bound to fail. But at the same time, he also argues that the bodily inaction of this thoroughly biopolitical self-realisation—for example, in the form of physical relocation, in his own case from China to the United States—can alter the body's situatedness in the intersectional grid of biopower to make available pathways for genuinely resistant thought and action, or as Zhang puts it in Foucault's words, 'a will for the otherwise'.

While the previous two sections have unpacked the biopolitical apparatus working through producing and making productive the body of the governed in its interrelated lines and logic of operation, the final section of the volume, entitled *Embodied Life: Erasure, Contagion, Immunisation*, is specifically dedicated to the biopolitical quality of the body that is not there—the body whose death must remain unwitnessed, sexual identity remain unrecognised, political and economic agency remain stalled not to interrupt the reproduction of the biopolitical One as the healthy, racially homogenous, unambiguously gendered and sexually normalised population to make live. Beginning with the latter dimension, Christian Klesse critically analyses the erasure of bisexual bodies from legal discourses and decisions which assert the right to asylum on the ground of prosecution for sexual orientation. Escaping and thereby undermining rigid, clearly delineated and visible lines of sexual normalisation, Klesse explores how bisexuality escapes established lines of legal justification and political support for asylum claims, rendering the status of bisexual asylum applicants deeply problematic.

While this could be remedied with the recognition of bisexuality as an identity 'label' which faces discrimination and prosecution, Klesse argues for caution here, highlighting the radical potentiality of bisexuality. Due to its intrinsic fluidity, bisexuality transcends established lines of biopolitical production and thus entails the potential to overcome the Western liberal idea of a naturally fixed, coherent and demonstrable identity as a necessary condition for political-legal protection. Contrary to this resistant escape from rigid lines of biopolitical production, Shubranshu Mishra's chapter begins with a governmentally enforced erasure of bodies. Mishra shows how the Indian government attempted to deprive political opposition in the Kashmir conflict of its material actualisation and visibility by making the bodies of resistant fighters disappear in anonymous mass graves dug and filled by coercively enlisted gravediggers. But, as suggested above, the political productivity of the body of the governed, the gravedigger employed by state forces, is not determined but open-ended. Providing accounts of personal conversations with Kashmiri gravediggers, Mishra reveals that their labouring bodies can constitute a source of quiet but physically manifest and directly effective resistance.

The gravediggers, who retained personal items from the killed fighters, buried them ceremonially and marked the sites, worked against the political erasure ordered by the government to not only resubjectify the *homines sacri* excluded for the reproduction of the political status quo but also retained the memory and hence the trajectory of their political opposition. The operationality of a biopolitical apparatus where exclusion functions constitutive for the political-economic status quo is also explored in Dimitra Kotouza's chapter. Kotouza emphasises the neoliberal underpinning and functionality of biopower and immunitarian governance. Analysing the Greek financial crisis of 2008 as biopolitical event where the intersecting lines of racialisation, gendering and sexual normalisation supported each other. Kotouza unpacks how biopolitical governance here socially enforces the patriarchal household model that in turn sustains the economic conditions of the crisis and the gendering, sexually normalising and racialising abjection it performs. Similarly to Bird, Kotouza thereby reconnects the biopolitical theories of Agamben and Esposito to the economic reproduction of the Body of capital which is marginalised in both.

Benoît Dillet's final chapter also unpacks political strategies of containment and immunisation in recent political and social crisis events. Dillet shows how the biopolitical functionality underlying contemporary political and economic production does not simply amount to a transition from a sovereign politics of death to a governance of life but rather combines both dimensions in an ecotechnics of biopower which constantly rewrites not only the category of life itself in its extent and quality but also the political rights and claims attached to it. Connecting the theories of sociopolitical immunisation developed by Esposito and Fréderic Neyrat, Dillet shows how the ecotechnological biopolitical machine produces exophobic communities. It works through and relies on the immunisation of political communities from an outside otherness, revealing the drive to remain untouched, to be distinct, as a third operational mode of biopower beyond the power to take lives and make live. Where Esposito remains vague and ultimately unconvincing on the possibility of an affirmative turn to rewrite othering biopolitics, Dillet shows how Neyrat offers a promising ecophilosophical perspective which seeks to reconceptualise the relationship between human political communities and their environment in a way which dissolves the immunitarian dialectic but without falling victim to the flat, de-differentiating ontology of capitalism, as new materialist approaches risk.

If the challenge for resistance to biopolitics 'is not globalisation as such, but "hominisation" as a process', as Dillet writes with reference to Neyrat, then a pathway for resistance might be the conceptualisation of a connective, not immunitarian, Anthropocenic biopolitics. Such a conceptual framework must grasp the relationship between politics and life, community and environment without recourse to the antipodes of difference and unity. On the contrary, it must make both critical-diagnostic and politically creative use of their productive relational middle ground. In the way it unpacks, challenges and resolves the binarisation of biopolitics—in the form of race and gender, gender and economy or sexuality—in thought and political practice, this collection is designed to stock such a conceptual toolbox which is worthy of contemporary biopolitical times. It seeks to both provide lenses suitable for their nuanced analysis and conceptual tools to sketch out theoretical and political alternatives.

NOTES

This collection is the productive development of papers presented and thoughts exchanged at the workshop 'The Body of the Governed: The Biopolitics of Race and Gender in Theory and Political Practice' which I co-organised with Iain MacKenzie at the University of Kent in December 2015. I would like to thank Dimitra Kotouza, who suggested the idea for this workshop and took the first initiative to make it happen, as well as the University of Kent for their generous support of this event. For their support and advice at many stages of developing and preparing this volume, I would further specifically like to thank Greg Bird, Marco Piasentier and Iain MacKenzie. Thank you as well to Dhara Snowden from Rowman & Littlefield International and the series editors Nicola Smith, Juanita Elias and Adrienne Roberts for their excitement about the collection, all of their guidance and—especially towards the end of the editing process—patience; and finally, of course, all of the chapter authors for their fantastic contributions and collaborative engagement which made this a—even though at times work-intensive—smooth and enjoyable editing process.

1. Unpacking the well-known ambiguity of 'sex' in Foucault which was pointed out by Judith Butler (1993), the biopolitical line of sex in the following is referred to as both gendering and sexual normalisation.

BIBLIOGRAPHY

Agamben, Giorgio. 2012. *The Kingdom and the Glory: For a Theological Genealogy of Economy and Government*. Palo Alto, CA: Stanford University Press.

Agamben, Giorgio. 2016. *The Use of Bodies*. Palo Alto, CA: Stanford University Press.

Althusser, Louis, ed. 2016. *Reading Capital*. London: Verso.

Apostolidou, Sofia, and Jules Sturm. 2016. "Weighing Posthumanism: Fatness and Contested Humanity." *Social Inclusion* 4 (4): 150–59.

Bachmann, Ingeborg. 2000. *Ich weiß keine bessere Welt. Unveröffentlichte Gedichte*. Munich: Piper.

Bachmann, Ingeborg. 2008. *Malina*. München: Süddeutsche Zeitung Bibliothek.

Balfour, Lawrie, Falguni Sheth, Heath Fogg Davis, Shatema Threadcraft and Jemima Repo. 2016. "Bodies in Politics." *Contemporary Political Theory* 15 (1): 80–118.

Balibar, Etienne. 1991. "Is There a 'Neo-Racism'?" In *Race, Nation, Class: Ambiguous Identities*, edited by Etienne Balibar and Immanuel Wallerstein, 17–28. London: Verso.

Butler, Judith. 1993. *Bodies That Matter: On the Discursive Limits of "Sex."* New York: Routledge.

Cheah, Pheng. 2013. "The Biopolitics of Recognition: Making Female Subjects of Globalization." *Boundary 2* 40 (2): 81–112.

Davis, Heath Fogg. 2012. "Sex-Classification Policies as Transgender Discrimination: An Intersectional Critique." *Perspectives on Politics* 12 (1): 45–60.

Deleuze, Gilles. 1992. "Postscript on the Societies of Control." *October* 59 (4): 3–7.

Deleuze, Gilles. 2004. *The Logic of Sense*. London: Continuum.

Deutscher, Penelope. 2010. "Reproductive Politics, Biopolitics and Auto-immunity: From Foucault to Esposito." *Journal of Bioethical Inquiry* 7 (2): 217–26.

Doran, Peter. 2017. *A Political Economy of Attention, Mindfulness and Consumerism: Reclaiming the Mindful Commons*. London: Routledge.

Esposito, Roberto. 2011. *Immunitas: The Protection and Negation of Life*. Cambridge: Polity.

Foucault, Michel. 1976. *Society Must Be Defended: Lectures at the Collège de France 1975–76*. New York: Picador.

Foucault, Michel. 1978. *The History of Sexuality I: An Introduction*. New York: Pantheon Books.

Foucault, Michel. 1988. "An Aesthetics of Existence." In *Politics, Philosophy, Culture: Interviews and Other Writings 1977–1984*, edited by L. D. Kritzman, 47–53. New York: Routledge.

Foucault, Michel. 1991. *Discipline and Punish*. London: Penguin Classics.

Foucault, Michel. 1997. "Sex, Power, and the Politics of Identity." In *Ethics, Subjectivity and Truth: Essential Works of Michel Foucault*, edited by Paul Rabinow, 163–73. New York: New Press.

Foucault, Michel. 2008. *The Birth of Biopolitics: Lectures at the Collège de France 1978–79*. Basingstoke, UK: Palgrave Macmillan.

Inhorn, Marcia Claire. 2009. *Reproductive Disruptions: Gender, Technology, and Biopolitics in the New Millennium*. New York: Berghahn Books.

Kantorowicz, Ernst Hartwig. 1997. *The King's Two Bodies: A Study in Mediaeval Political Theology*. Princeton, NJ: Princeton University Press.

Kelly, Mark G. E. 2015. *Biopolitical Imperialism*. London: Zero.

Larsen, Lars Thorup. 2011. "The Birth of Lifestyle Politics: The Biopolitical Management of Lifestyle Diseases in the United States and Denmark." In *Governmentality Current Issues and Future Challenges*, edited by Ulrich Bröckling, Susanne Krasman and Thomas Lemke, 201–24. Milton Park, UK: Routledge.

Lazzarato, Maurizio. 2002. "From Biopower to Biopolitics." *Pli: The Warwick Journal of Philosophy* 13: 99–110.

Lecat-Deschamps, Jean-Amos. 2015. "Vers une datapolitique?" *La Deleuziana* 1: 141–57.

Lemm, Vanessa, and Miguel Vatter, eds. 2014. *The Government of Life: Foucault, Biopolitics, and Neoliberalism*. New York: Fordham University Press.

Lobo-Guerrero, Luis. 2016. *Insuring Life: Value, Security and Risk*. London: Routledge.

Marx, Karl. 1963. *The Poverty of Philosophy*. New York: International Publishers.

Marx, Karl. 1976. *Capital: A Critique of Political Economy*, vol. 1. Harmondsworth, UK: Penguin.

Milchman, Alan, and Alan Rosenberg. 2011. "Michel Foucault: An Ethical Politics of Care of Self and Others." In *Political Philosophy in the Twentieth Century: Authors and Arguments*, edited by Catherine H. Zuckert, 228–37. Cambridge: Cambridge University Press.

Mitchell, David, and Sharon Snyder. 2015. *The Biopolitics of Disability: Neoliberalism, Ablenationalism, and Peripheral Embodiment*. Ann Arbor: University of Michigan Press.

Mitchell, David, and Sharon Snyder. 2016. "The Matter of Disability." *Journal of Bioethical Enquiry* 13 (4): 487–92.

Pooja, Rangan, and Rey Chow. 2013. "Race, Racism, and Postcoloniality." In *The Oxford Handbook of Postcolonial Studies*, edited by Graham Huggan, 1–19. Oxford: Oxford University Press.

Prozorov, Sergej. 2017. "Living à la mode." *Philosophy & Social Criticism* 43 (2): 144–63.

Repo, Jemima. 2015. *The Biopolitics of Gender*. Oxford: Oxford University Press.

Repo, Jemima. 2018. "The Life Function: The Biopolitics of Sexuality and Race Revisited." In *Biopolitical Governance: Race, Gender, Economy*, edited by Hannah Richter, 41–57. London: Rowman & Littlefield International.

Rouvroy, Antionette. 2011. "Governmentality in an Age of Autonomic Computing: Technology, Virtuality and Utopia." In *The Philosophy of Law Meets the Philosophy of Technology: Autonomic Computing and Transformations of Human Agency*, edited by Mireille Hildebrandt and Antoinette Rouvroy, 119–40. Milton Park, UK: Routledge.

Rouvroy, Antionette. 2016. "Algorithmic Governmentality: Radicalisation and Immune Strategy of Capitalism and Neoliberalism?" *La Deleuziana* 3: 30–36.

Smith, Sara, and Pavithra Vasudevan. 2017. "Race, Biopolitics, and the Future: Introduction to the Special Section." *Environment and Planning* 35 (2): 210–21.

Stoler, Ann Laura. 1995. *Race and the Education of Desire: Foucault's History of Sexuality and the Colonial Order of Things*. Durham, NC: Duke University Press.

Tarizzo, Davide. 2011. "Biopolitics and the Ideology of 'Mental Health'." *Filozofski Vestnik* 32 (2): 135–49.

Tremain, Shelley. 2015. *Foucault and the Government of Disability*, 2nd ed. Ann Arbor: University of Michigan Press.

Vatter, Miguel. 2014. *The Republic of the Living: Biopolitics and the Critique of Civil Society*. New York: Fordham University Press.

Weheliye, Alexander G. 2014. *Habeas Viscus: Racializing Assemblages, Biopolitics, and Black Feminist Theories of the Human*. Durham, NC: Duke University Press.

Welch, Rhiannon Noel. 2016. *Vital Subjects: Race and Biopolitics in Italy, 1860–1920*. Liverpool: Liverpool University Press.

Part I

THE POLITICS OF LIFE BEYOND FOUCAULT

Chapter 1

Foucault and the Two Approaches to Biopolitics

Marco Piasentier

What is biopolitics? What kind of relationship does biopolitics establish between politics and biology? Although the etymology of the term 'biopolitics' seems to suggest a straightforward meaning resulting from the relationship between biological life and politics, the current literature is characterised by a wide variety of definitions. As the social theorist Thomas Lemke notes in his thoughtful introduction to this field of research, '[p]lural and divergent meanings are undoubtedly evoked when people refer to biopolitics' (Lemke 2011, 2). Lemke is not the only scholar to acknowledge the difficulty in establishing a satisfying definition of this term; the scale of the problem is well exemplified by the decision of the philosopher Roberto Esposito to begin his major work on the topic with a chapter entitled 'The Enigma of Biopolitics'. In the first chapter of *Bios: Philosophy and Biopolitics*, Esposito traces the enigmatic character of biopolitics back to the thinker who introduced it into the Continental debate—Michel Foucault—and maintains that the problematicity encountered by the French philosopher concerns 'the same logical and semantic configuration' (Esposito 2006, 43) of biopolitics. According to Esposito, the impasse characterising this field of research depends on the fact that 'notwithstanding the theorization of their reciprocal implication . . . politics and life remain indefinite in profile and in qualification' (Esposito 2006, 43–44).

In this chapter I focus on the work of Foucault, with the aim of explaining the impasse in defining the notion of biopolitics. Following Esposito, I claim that it is the lack of a correct articulation of the relationship between politics and life that lies at the core of the 'enigma of biopolitics'. However, the enigma does not concern the lack of inquiry into the two terms comprising this concept; at stake is a deeper and more complicated issue. I argue that when politics and biological life meet to constitute the notion of biopolitics, their function becomes that of representing two opposing theories of the human being. In turn, these two ways of defining the human determine two mutually exclusive approaches to biopolitics: the discursive and the vitalist. The 'enigma of biopolitics' is the name of this fracture.

The two definitions of the human being that I take into consideration in this chapter contribute, from opposite sides, to challenging his ultramundane origin. The reason for their conflict is the definition of the world to which the human being is returned after his fall. On the one hand, the notion of the human being of the discursive approach to biopolitics can be summed up, paraphrasing a famous Foucauldian passage, with the sentence: the human being is 'a thing of this world'. The original Foucauldian passage reads as follows:

> Truth is a thing of this world: it is produced only by virtue of multiple forms of constraint and it introduces regular effects of power. Each society has its regime of truth, its 'general politics' of truth—that is, the type of discourse it accepts and makes function as true; the mechanisms and instances that enable one distinguish true and false statements. (Foucault 2000a, 131)

Paraphrasing this quotation and affirming that the human being is 'a thing of this world' means that every society has a general politics of the human being—that is, the mechanisms and instances that enable every society to distinguish between normal and abnormal. Assuming this mundane definition of man means affirming that there is no human being prior to the '*dispositifs*' or historical 'machinations'—to use a Heideggerian term—so the human being has always already been thrown into the processes of subjectification and desubjectification. Reaching a point where it is possible to say what the human being is remains structurally impossible.

On the other hand, the definition of the human being proper to the vitalist approach to biopolitics entails a form of animalisation of the human being where his essence is found in the natural world of life. This definition of man can be expressed in Foucauldian terms, paraphrasing another passage on truth. The ultramundane notion of truth is no longer returned to the mechanisms of discourse and power but to the horizon of biological life: 'The true\false dichotomy and the value accorded to truth constitute the most singular way of living that has been invented by a life that, from the depths of its origin, bore the potential for error within itself' (Foucault 2000b, 477). In this case the human being is not the product of social practices but is the result of the natural evolution of life. The human being becomes an animal among other animals. Social practices and their true or false value are nothing more than an invention of a life which seems to be animated by a force aimed at its own survival and reproduction.

The opposition between these two notions of man determines two approaches to biopolitics. They can be distinguished from one another precisely because of the hierarchy they impose on the two terms comprising the word 'biopolitics': biological life and politics. Whereas the discursive approach implies the politicisation of biology, with the consequent reduction of every biological definition of the human being to an epiphenomenon of political struggle, the vitalist approach involves the biologisation of politics with the consequent reduction of politics to an epiphenomenon of biological life.

The failure to articulate a nonreductionist relationship between life and politics implies a teleology which inevitably informs both Foucauldian approaches to biopolitics. Beyond the antithesis between the discursive and the vitalist human being, it is, therefore, possible to find a common ground characterised by an unwanted teleological drift. This teleology is a force intended as the site of permanent openness and resignificability, an excess always overflowing its forms, which expresses a constant activity aimed at a work of contestation and metamorphosis set against fossilisation in a fixed identity. This teleology—which prescribes to be like wanderers who never belong to a place but belong to travel itself—can be defined in Foucauldian terms as an 'impatience for liberty' (Foucault 2000a, 319), in the case of the discursive approach and, paraphrasing Foucault, as an impatience for life, in the vitalist approach.

In this chapter, I show how these two definitions of the human being and consequent approaches to biopolitics coexist in the work of Foucault. It is precisely because of this conflict secretly operating in Foucauldian thought that his definition of biopolitics remains enigmatic. This vagueness should not be interpreted as a weakness but instead as a clear sign of Foucault's great capacity for grasping and describing the complexity of the theories involved in biopolitics. In the following pages, I do not propose a totalising explanation of Foucauldian thought, but I try to reveal how his work has become the anchoring point in defining two perspectives that play a hegemonic role in the contemporary understating of the relation between biological life and politics. The fracture between these two approaches undermines the possibility offered by biopolitics to sit astride the wall that today divides naturalism and critical theory, biology and politics. The purpose of this chapter is not to offer a hasty alternative but rather to explore the extent of this fracture in order to lay an adequate foundation for a new biopolitical inquiry.

I begin by addressing the discursive approach; later, I consider the vitalist one. Finally, I show how the fracture between these two conceptions of the human being defines an ontological limbo between the ultramundane and a truly mundane theory of the human being.

MAN IS DEAD, LONG LIVE MAN

At the end of *The Order of Things*, Foucault proclaims the death of man, arguing that 'man is an invention of recent date. And one perhaps nearing its end . . . like a face drawn in sand at the edge of the sea' (Foucault 2005, 422). In a 1978 interview, he goes back to that famous passage to further clarify its meaning:

> In *The Order of Things*, I made the mistake of presenting this death as something that was under way in our era . . . in the course of their history, men have never ceased to construct themselves, that is, to continually displace their subjectivity, to constitute themselves in an infinite, multiple series of different subjectivities that will never have an end and never bring us in the presence of something that would be 'man'. (Foucault 2000b, 275–76)

The interplay between sea and sand described in the two quotations defines a process of subjectification and desubjectification able to continually shape new human figures. The death of man should not to be seen as man's definitive disappearance but, rather, as a way of interrogating his construction and deconstruction through the inquiry into those discourses giving form to and erasing his faceless face. The process described in the two quotations has both a diachronic and synchronic dimension: every subjectification is built upon the ruins of a previous one, and its constitution traces the geometries of inclusion/exclusion proper to each society, such as 'the relationship between sanity and insanity, or sickness and health, or crime and the law' (Foucault 2000a, 318). Critical inquiry shows that the material constituting each of these theories of the human—and, consequently, the perimeters of exclusion they establish—is not the rock of a necessary metaphysics but the sands of a contingent history.

The investigation into the forms of subjectification and desubjectification ultimately demonstrates that man 'is a thing of this world' (Foucault 2000b, 131). He is produced—to continue the paraphrase of the famous Foucauldian sentence on truth—only by virtue of multiple forms of constraint; each society has its 'general politics' of man—that is, the types of discourses which constitute man and provide the means to distinguish between normal and abnormal (see Foucault 2000b, 131). If maintaining that the human being 'is a thing of this world' is to claim that there is no human being prior to the processes of social construction, it follows that the term 'world' cannot help but resonate with the legacy of Heidegger. Foucault himself acknowledges the importance of the German philosopher in an interview conducted in 1984, when he declares that 'for me Heidegger has always been the essential philosopher. . . . My entire philosophical development was determined by my reading of Heidegger' (Foucault 1988, 250). In particular, it is the influence of the later Heidegger that emerges in the works of Foucault.

As Gadamer has already noticed, 'Heidegger himself, after the "turning", abandoned his transcendental conception of self' (Gadamer 1981, 104). Relinquishing once and for all the idea of the human being having a permanent structure, the later Heidegger affirms that the taking place of the human being is always already the 'unveiling' of the 'being thrown' and no essence of the human being prior to the 'being-in-the-world' can be posed. Whereas in *Being and Time* 'anxiety and anticipation of death revealed structural aspects of Dasein that couldn't be defined wholly in terms of the world' (Braver 2007, 218), with 'the turn' (*die Kehre*), Heidegger fulfils *Being and Time*'s conception of 'being-in-the-world', marking a 'turn' not only in his thought but in the history of Continental philosophy itself. The human being has to be intended to be so intrinsically united to the act of existing that using the verb 'to have' in describing the relation between human being and his 'being-in-the-world' is incorrect, since it creates the impression of a fracture between essence of the human being and his existence (see Agamben 2016). The essence of the human being does not have 'being-in-the-world' as its mode, but 'the essence of the human being consists in being-in-the-world' (Heidegger 1998, 266). 'Being-in-the-world' designates the essence of the human being in the sense that there is no world if there is no man, and man exists, in turn, only in the form of 'being-in-the-world'.

In the later Heidegger, as well as in Foucault, the human being is always already consigned to his existence in the world: there is no human being beyond existence so that the act of existing brings the essence of the human being into being. The critical inquiry proposed by Foucault—intended as a 'work on our limits, that is, a patient labor giving form to our impatience for liberty' (Foucault 2000a, 319)—is precisely a way of showing how every understanding of ourselves is kept within what Heidegger calls 'epochs' and Foucault 'ontologies of ourselves'—that is, one of the many definitions of the human resulting from the interplay between sea and sand and, in other words, one of the many ways in which the 'being-in-the-world' of the human being designates his essence. Whereas I later take into consideration the topological position of that human being who describes the interplay between sea and sand, I now address the relation between Foucault's notion of subjectification and biopolitics.

ANIMALISATION

In the final section of the *History of Sexuality*, volume 1, Foucault introduces the notion of biopolitics. In line with the aim of his critical inquiry, addressing this new field of research consists in individuating and describing the emergence of a new subjectification of the human being. Foucault maintains that 'for millennia, man remained what he was for Aristotle: a living animal with the additional capacity for a political existence; modern man is an animal whose politics places his existence as a living being in question' (Foucault 1978, 143). The interplay between sea and sand that leads to 'biological modernity' dissolves a form of subjectification that—despite the many and different interpretations—has had a hegemonic role in Western society since ancient Greek philosophy, and that finds one of its most emblematic examples in the Aristotelian definition of the human being. The desubjectification and the consequent possibility of a subjectification enacted by 'biological modernity' entails what Foucault defines as an 'animalization of man' (Foucault in Dreyfus and Rabinow 1983, 138). The key characteristic of this process of animalisation is a redefinition of the hierarchical structure between mind and body which has regulated most of Western thought for millennia. Although Aristotle's mind/body problem is open to different and even contrasting interpretations, it is possible to affirm without a great margin of doubt that his idea of the human being is informed by that 'meta' that allows the overcoming of animal *physis* in the direction of a transcendent dimension.

In the first book of the *Nicomachean Ethics*, Aristotle states that the highest good of human beings coincides with happiness (*εὐδαιμονίαν*); the ultimate task of politics should be that of reaching the highest good, and therefore happiness. The possibility of accomplishing the task of politics rests on the understanding of what happiness is for human beings; in turn, knowing what happiness is requires a definition of the human being, and Aristotle provides this definition, especially in the *De Anima* and *Metaphysics*. In the second book of the *De Anima*, Aristotle develops a hylomorphic (*ὕλη*, 'matter', and *μορφή*, 'form') doctrine whereby the soul is the form of the body, and the body the matter of the soul. The soul is a general principle of life which does not merely belong to living beings having a mind but can also be found in plants and animals. The soul bears the same relation to the body that, for example, seeing bears to the eye: 'the eye is the matter for sight, and if this fails it is no longer an eye, except homonymous, just like an eye in stone or a painted eye' (Aristotle 2002, 10). It follows that 'we should not ask whether the wax and the impression are one, any more than whether the matter of each thing and that of which it is the matter are one' (Aristotle 2002, 9). However, this hylomorphism is complicated by a form of dualism between the human body and its mind. Aristotle divides the soul into three parts: the nutritive soul proper to plants, the sensitive soul proper to all animals and the rational soul of human beings. In the *De Anima*, he also writes that the rational soul 'seems to be a different kind of soul, and this alone can exist separately as the everlasting can from the perishable' (Aristotle 2002,

13). Therefore, 'this intellect is distinct . . . and this alone is immortal and eternal' (Aristotle 2002, 60).

With the crossing of 'the threshold of biological modernity', which Foucault roughly places in the eighteenth century, 'modern western societies took on board the fundamental biological fact that human beings are a species' (Foucault 2007, 16). The human life and its mechanisms are brought into the realm of political calculation thanks to a new definition of human being, according to which the explanation of the human being is exhausted by nature, containing nothing 'supernatural'. The process of desubjectification determined by the disappearance of the transcendent soul does not lead to the 'death of men' but to a new form of subjectification, according to which human beings conceive of themselves as the more or less accidental result of a natural history that holds together all living beings and therefore makes an animal among other animals of the human being.

The peculiarity of the human species lies in the intertwining of history and biology, nurture and nature. Depending on the perspective from which one sees this knot, one tends to put more weight on the cultural or the biological side. In the former case, one tends to see the evolution of the human being as a lucky accident, to consider his innate traits as being not prescriptive but always subject to being reshaped by cultural practice. In the latter case, one infuses progressivism and determinism into the evolutionary story and considers the human being as a set of innate traits with specific inviolable functions. Despite the fact that the choice may lean more to the cultural or the biological side, one still finds the human being within that subjectification which Foucault defines as the modern 'animalization of man' and which serves as the premise for 'both the development of the possibilities of the human and social sciences, and the simultaneous possibility of protecting life and the holocaust' (Foucault in Dreyfus and Rabinow 1983, 138).

OUTSIDE

The Foucauldian critical inquiry on biopolitics leads one to ask whether his analysis accepts as true the definition of man as a biological being or whether the biological explanation of the human remains—from the perspective of Foucault's critical inquiry—nothing more than a fiction. Providing an answer to this question means mapping out the relation between Foucault's critical inquiry as previously examined and the subject of his inquiries. In the introduction to *The Archeology of Knowledge*, Foucault himself provides an answer to this topological question by writing that 'no, I'm not where *you* are lying in wait for me, but over here, *laughing* at *you*' (Foucault 1972, 17). The sense of ungraspability conveyed by his words is not a rhetorical expedient but is rooted in the very philosophical premise of his critical inquiry. If every 'ontology of actuality' is always only the historical result of the endless process of subjectification and desubjectification, none of these definitions of the human being can benefit from an epistemological privilege over the others. The task of critical inquiry is to dissolve the chimera of the origin by allowing its nonoriginary character to emerge. Foucault's work shows that every construction and deconstruction of man rests on an 'originary in/origin'[1] which makes it impossible for each subjectification and desubjectification to impose itself in its complete presence or absence.

Wondering whether Foucault finds in biology a preferable theory to define the human being is equivalent to asking about the position from which he speaks. In both cases it will ultimately find that the continuous work of contestation of critical inquiry requires that Foucault's position, as well as his definition of the human being, be that of 'an outer bound where it must continually contest itself' (Foucault 1994, 152). When the critical inquiry of man 'arrives at its own edge, what it finds is not a positivity that contradicts it but the void that will efface it. Into that void it must go, consenting to come undone in the rumbling, in the immediate negation of what it says, in a silence that is not the intimacy of a secret but a pure outside where words endlessly unravel' (Foucault 1994, 152). The 'pure outside' is a place where one does not 'discover what he is [we are]' but which forces him to the endless obligation 'to refute what he is [we are]' (Foucault in Dreyfus and Rabinow 1983, 216). The position from which Foucault is laughing is, therefore, a point of flight always overflowing its forms. The outside does not refer to any specific 'ontology of actuality'; its only property is to mark the impossibility of offering a definition of the human being escaping the never-ending process of subjectification and desubjectification. It can be argued, with Foucault, that the endless unravelling of words—namely, the endless possibilities of defining the human being—marks the silence characterising an ultimate definition of the human being and, with Heidegger, highlights that the only authentic discourse on the human being is silence. But if there are no words to define the human being—if, in other terms, it will never be possible to find oneself in 'the presence of something that would be "man"' (Foucault 2000b, 275–76)—how should one consider the discourse which nowadays maintains that the human being is an animal species resulting from the evolution of life? This question finds a clear answer in the debate on human nature between Chomsky and Foucault.

In discussing the existence of some innate traits of mind and human nature, Foucault disagrees with Chomsky 'when he places the principle of these regularities, in a way, in the interior of the mind or of human nature', since 'to say that these regularities are connected, as conditions of existence, to the human mind or its nature, is difficult for me to accept: it seems to me that one must, before reaching that point . . . replace it in the field of other human practices, such as economics, technology, politics, sociology, which can serve them as conditions of formation, of models, of place, of apparition, etc.' (Chomsky and Foucault 2006, 29). But the point where it is possible to define 'the interior of the mind or of human nature' is intrinsically unreachable, since the Foucauldian 'impatience for liberty' forces one to a constant refusal of every definition of the human being.

The Foucauldian method challenges the normative consequences that deterministic biology has on issues such as gender, health, race and social class systems, but only by reducing every definition of biology to a discursive construction. It is not obviously Foucault's aim to deny biology itself, and in the dialogue with Chomsky, it is possible to see the tension characterising his position. Despite Foucault trying to make sense of the ambivalent role of biology in his theory, it eventually becomes clear that the fictional character of every definition of the human being as biological entity is an inevitable side effect intrinsic to his critical inquiry. Foucault cannot ultimately accept the existence of some innate principles characterising the human being even when Chomsky claims that, 'as a matter of biological and anthropological fact, the nature of human intelligence certainly has not changed in any substantial way, at least since the seventeenth century, or probably since Cro-Magnon man' (Chomsky and Foucault 2006, 33). The further clarification proposed by Chomsky does not change Foucault's position; he insists on the constructive essence of every definition of the biology of the human being by affirming that they have to be considered an invention emerging 'from our society, from our civilisation, from our culture' (Chomsky and Foucault 2006, 43).

It is evident that the point of fracture between the two speakers is not about the impact of culture in shaping human nature; it is far more radical. Beyond every articulation of the relationship between culture and biology, nurture and nature, every time one tries to trace Foucault's critical inquiry back to a biological definition of the human being, even those most open to the influence of culture modelling and transforming biology, one is depriving this method of its novelty and force. From the perspective of critical inquiry, as Agamben clearly states it in *The Open*, '*Homo sapiens*, then, is neither a clearly defined species nor a substance; it is, rather, a machine or device for producing the recognition of the human' (Agamben 2004, 26). *Homo sapiens* is nothing but one of many politics of the human resulting from the functioning of the 'anthropological machine'. Instead of looking for ever more sophisticated modes of articulating the relation between human beings and biology, the task of critical inquiry has to go in the opposite direction and show the 'central emptiness' characterising the human being, and to risk himself in this void (see Agamben 2004, 92).

The emptiness of which Agamben speaks seems to be another name for Foucault's 'outside'. When one risks oneself in this emptiness, which coincides with the silence marking the lack of identity, one has to conclude that *Homo sapiens* is nothing but 'an invention of recent date' (Foucault 2005, 422). It follows that affirming that the *human being is a thing of this world*—namely, that the essence of the human being coincides with his existence and therefore that there is no human being before or beyond his thrownness—leads to the inevitable conclusion that the idea of the human being as one animal among other animals is nothing but one of the many ways in which human beings have constructed themselves. Every biopolitical theory that moves from these premises might also neutralise the negative political outcomes of a deterministic biology, but the price to be paid is very high. If the aim of the discursive approach to biopolitics is that of neutralising the logic of presupposition informing biopolitics, in reality the result of this attempt leads to deny the biological conception of the human being but saving the logic of presupposition, which becomes the presupposition of the 'impatience for liberty': the endless contestation of every definition of the human being guided by the awareness that the nameless name of man presupposes an originary silence from which the infinite ways of naming him arise.

ANOTHER FOUCAULT

In the final section of *Homo sacer* there is a comment which proves very interesting when compared with Foucault's approach to biopolitics as presented above. Agamben notes that:

> at the end of the first volume of the *History of Sexuality,* having distanced himself from the sex and sexuality in which modernity, caught in nothing other than a deployment of power, believed it would find its own secret and liberation, Foucault alludes to a 'different economy of bodies and pleasures' as a possible horizon for a different politics. The conclusions of our study force us to be more cautious. Like the concepts of sex and sexuality, the concept of the 'body' too is always already caught in a deployment of power. (Agamben 1998, 104)

Agamben claims that Foucault is placing the natural body at the heart of a political proposal, endorsing, in this way, the very biopolitical ground he has so often put into question. Agamben is not the only prominent reader of Foucault to notice such a discrepancy.

In a 1989 article entitled 'Foucault and the Paradox of Bodily Inscriptions', Judith Butler maintains that the idea of the body as constructed by regimes of discourse/power is deeply indebted to the thought of Foucault; however, in some important theoretical essays and writings he presents the body as being ontologically prior to the process of construction. In works such as *The History of Sexuality*, volume 1, the essay 'Nietzsche, History, and Genealogy', and *Discipline and Punish*, Butler detects the malfunctioning of the discourse analysis and individuates the cause of this fault in the idea of inscription. Foucault claims that regimes of discourse and power 'inscribe' themselves into the body. Maintaining that something is inscribed into something else implies that the latter element pre-exists the former. The process of inscription, therefore, suggests that the body is ontologically pregiven to the mechanism of cultural construction.

Butler continues by demonstrating that, due to the problematic use of the notion of inscription, the critique Foucault directs towards Nietzsche and Freud can be turned against himself. According to Foucault, both Nietzsche and Freud would assume a prediscursive ontology of the body and its drives, and this ontological assumption would be the ground for their 'repressive hypothesis',[2] which, despite the differences between them, would entail that culture is a means of repression of the most vital drives of the body. Butler claims that in Foucault it is possible to find a residue of this conception of the prediscursive vitalist body, precisely the same as Agamben has detected, raising 'the question of whether there is in fact a body which is external to its construction, invariant in some of its structures, and which, in fact, represents a dynamic locus of resistance to culture per se' (Butler 1989, 602).

If one thinks for a moment about the debate between Chomsky and Foucault as just analysed, it appears without doubt that, as Butler writes, in the debate 'Foucault wants to argue—and does claim—that bodies are constituted within a specific nexus of culture or discourse/power regimes, and that there is no materiality or ontological independence of the body outside of any one of those specific regimes' (Butler 1989, 602). As I show in the following sections, it is also correct to maintain that in some texts by Foucault, 'the body and its forces' (Foucault 1977, 25) is ontologically prior to the process of construction. Both points elaborated by Butler are very well supported by textual evidence; what should be placed in question is her attempt at fitting Foucault's notion of the body—as elaborated in the texts she quotes—within the discursive approach.[3] Foucault's philosophy is complex, and cannot be reduced to a single, monolithic ontology; for this reason I argue that the Foucauldian definition of the body that Butler is taking into consideration has nothing to do with the discursive approach or with the position Foucault takes in the debate with Chomsky, but it belongs to a different ontology, the vitalist one.

THE IMPATIENCE FOR LIFE

The attempt to force Foucault's prediscursive notion of the body into the discursive approach is destined to fail. It is not a matter of correcting the imperfections of the Foucauldian discursive approach but rather of venturing towards a different philosophical horizon whose grounding principle is not that of the 'impatience for liberty'. Foucault is no longer the critic of that process of animalisation proper of biopolitics but embraces the animalisation of man as an ontological starting point from which to develop his affirmative politics of life. The world that serves as an ontological fore-structure of the human is no longer that of the historical regimes of power and discourse but that of nature and biological life. However, the process of animalisation envisioned by Foucault infuses into the materiality of the living body an *élan vital* that imposes a new, natural destiny for the human being. This is precisely the point where the animalisation of the human being—intended as that explanation able to make sense of the biological origin of the human being without introducing any form of anthropomorphic and teleological drifts—fails. The vitalist approach to biopolitics, which I present in the final pages of this chapter, is grounded in this anthropomorphic and teleological theory of the living, which is not antithetical to current evolutionary biology but in line with a certain interpretation of it.

A possible point of entry into the Foucauldian philosophy of life is the Latin term *error-oris*, in its double meaning of error and errancy.[4] Foucault introduces this concept in the essay he dedicates to Canguilhem, where he proposes an account of biological life as being that which 'is destined to "err" and to be "wrong"' (Foucault 1994, 476). Biological life is not often associated with terms such as wandering and anomaly, but rather with progress and optimality. The contemporary hegemony of a deterministic interpretation of the theory of evolution by natural selection has depended on the influence played by the so-called adaptationist paradigm in the field of biology ever since Darwin. This approach infuses life with a functional determinism according to which organisms are elegant machines whose traits are designed to correctly perform their proper function. In turn, evolution becomes a linear and progressive path towards the improvement of the functionality of each biological structure. The philosophy of *error* challenges the one-to-one relationship between form and function. Evolution stops being the 'just-so story' of the elegant and efficient development of biological design and becomes a clumsy path whereby most of the traits are kludges co-opted to serve different functions in the course of their natural history. It follows that 'the cause of the genesis of a thing and its final usefulness, its actual employment and integration into a system of purposes, lie *toto caelo* apart' (Nietzsche 2011, 50–51), or, to put it in more contemporary biological terms, 'any organ, during its history, undergoes a series of quirky shifts in function' (Gould 2002, 1217).

The demonstration that the origin of a trait usually differs from its current function is only the first step towards establishing 'the ground of contingency and unpredictability in history' (Gould 2002, 1217). The philosophy of *error* also demonstrates that traits with no obvious function or that are even dysfunctional can be co-opted to serve new adaptive functions. The popular example of the panda's 'thumb', wonderfully narrated by the palaeontologist Stephen J. Gould (1980), can be used to provide scientific evidence of precisely that Nietzschean idea of 'ennoblement through degeneration' which lies at the very core of the Foucauldian notion of life as *error* (see Nietzsche 1996, 107). The giant panda has on its front paws what seems to be a thumb which it can press against its palm pad. The thumb is not one of the normal digits but is in fact a single bone resulting from an enlarged extension of the radial sesamoid. Thanks to this malformation giant pandas can hold and manipulate bamboo—the staple food of their diet—by passing the stalks between the 'thumb' and the remaining digits. As Foucault himself would probably say, contrary to the adaptationist approach, the example demonstrates how 'the question of anomaly permeates the whole of biology. And it must also be asked to account for the mutations and evolutive processes to which they lead' (Foucault 1994, 476). The possibility of *error* is not something to purge from an ideal portrayal of the evolution of life, but it needs to be taken into account as one of the most important sources of evolutionary change.

Despite the critique of the normative ground of adaptationism, the philosophy of life as *error* does not completely challenge the concept of biological normativity but rather allows its core to emerge. The ground zero of normativity of life is an undeconstructable force, able to construct and deconstruct the functionality of the different traits of an organism. Life is a creative force able to use adaptive and nonadaptive traits in subversive ways by displacing them from the function in which they have been hitherto deployed. In order to fully operate, the creative force of life has to challenge the possibility of a fixed function for a trait because this would render life incapable of establishing new functions to better reproduce and survive. In Foucault's words: '"error" constitutes not a neglect or a delay of the promised fulfilment but the dimension peculiar to the life of human beings and indispensable to the duration *[temps]* of the species' (Foucault 1994, 476). In the philosophy of life as *error*, evolution becomes a clumsy path whereby 'natural selection operates as the creative force of evolutionary change' (Gould 2002, 139). However, the philosophy of life as *error* does not completely dispense with the intrinsic normativity of life since the latter becomes a rebellious force able to resignify even anomalies into vital possibilities for survival and reproduction. Paraphrasing the Foucauldian expression 'impatience for liberty', I argue that, in the vitalist Foucault, life is characterised as an impatience for movement and transformation as, in other words, an *impatience for life*.

Despite the crucial importance of the philosophy of *error* in challenging the adaptationist paradigm and offering a 'pluralist' reading of the theory of evolution by natural selection, it is nevertheless necessary to question the vitalist residue still inhabiting it. This critique of the philosophy of *error* is coherent both with the critique of biological determinism and also with Darwin's original idea of life. As one reads in Darwin's autobiography: 'there seems to be no more design in the variability of organic beings and in the action of natural selection, than in the course which the wind blows' (Darwin 2002, 50), which means that biological structures simply operate and one should avoid projecting normativity into them. Introducing the term *error* to characterise life itself is a dangerous move since it inevitably implies the idea of correctness, as if there were a right and wrong way for life itself to absolve its intrinsic purpose, but life has no intrinsic purpose, not even the one of survival and reproduction. Only by liberating life from every intrinsic finalism is it possible to claim that 'human beings are not for anything. Nor are [our] their eyes of the flora that thrive within them [us]' (Cummins and Roth 2009, 81). Life itself has no intrinsic meaning and therefore by itself cannot provide direction as to how human beings should live and organise their individual and political existence. This should be the presupposition of every affirmative biopolitics, because only if human beings find themselves brought into life without the vehicle of a supernatural or natural purpose can space for an affirmative biopolitical realm be opened fully. But this is not the ontological presupposition shared by many contemporary biopolitical theorists.[5] On the contrary, they move from a vitalist conception of biological life and build up their biopolitics starting from an anthropomorphic and teleological *élan vital*. The result is a biopolitics of *error*.

BIOPOLITICS OF *ERROR*

The relation between biological life and politics proper of the vitalist approach to biopolitics has not to be understood only within the terms set by the repressive hypothesis. In the vitalist approach, life is not simply that counterhegemonic creative force able to contrast the orthopaedics of disciplinary techniques trying to annihilate the man by forcing him to conform to a fixed script of social identities. The horizon within which the idea of life has been developed is far more articulated and cannot be reduced to the contrast between the unmanageable and abnormal freedom of life and the managerial and normalising apparatus of society. According to the vitalist approach, the repressive hypothesis is only one of the many possibilities for envisioning the relationship between biological life and culture. The power of life is not a simple activity of resistance but becomes the very agent of creation of culture itself. Culture is—as it clearly emerges in the text that Foucault dedicates to

Canguilhem—one of the many 'inventions' of the creative power of life, and only by starting from this premise can the repressive hypothesis be developed.

The biological structures serving the development of culture can be the direct result of adaptations evolved over the course of natural history or they can be a side-effect of some other structures. What really matters is that human culture is first and foremost understood and evaluated in the terms by which it affects the success and survival of individuals and groups. Despite the evolutionary history of the emergence of culture, what really matters is why culture has become at a certain point necessary to the survival of the human being or, as Foucault himself writes, 'indispensable to the duration [*temps*] of the species' (Foucault 1994, 476). The conception of the human being as natural species, namely as *Homo sapiens*, is therefore no longer a cultural construction, as in the discursive approach, but what humans really are; culture, instead, becomes instrumental to the survival of the human species. Only at this point can the repressive hypothesis enter the scene; that is, the question about those forms of culture which, by denying the existence of life or proposing an inaccurate concept of it, end up developing social and institutional structures that repress the creative energy of life itself. It is against this background that we can understand the Foucauldian vitalist project of a 'different economy of bodies and pleasures' (Foucault 1978, 159).

The famous passage from *The History of Sexuality* has been the point of departure for imagining a culture and therefore a politics which 'could no longer be a culture of biological determinism, could no longer be, in other words, a culture against nature' (Malabou 2008, 30). The apparent groundless ground defined by the equation between freedom and life has become the mantra for those advocating for another politics or, better, a biopolitics of *error*, of contamination, of the multitude in opposition to that biological determinism which has produced a damaging politics of security and control. Because of its potential to resist any fossilisation into a politics of identity, the biopolitics of *error* can open up a new horizon where the 'power over life' is apparently defeated by the 'ungovernable' and 'rhizomatic' 'power of life'. The biopolitics of *error* is aimed at avoiding that 'politics over life' determined by the need to defend a supposed purity from contamination and diversity. In opposition to the autoconservative syndrome of those who consider the body as a static and closed unity, the biopolitics of *error* finds that 'diversity, alterity, hybridisation are not necessarily a limit and a danger from which we have to defend ourselves in the name of a self-centred purity of the individual and the species' (Esposito 2011, 83–84).

On the contrary, ideas of error and diversity become the means by which to imagine a different concept of biological life in which immunisation can be reached also by contamination and adaptation thanks to maladaptations, as in the example of the panda. Despite the biopolitics of *error* challenging the binary biopolitics of a deterministic notion of life, it presents a problematic normative ground. The aspiration for a politics of the living in opposition to the 'thanotopolitics' of the nineteenth and twentieth centuries has been realised thanks to a new conception of life, no longer intended as a synonym for functional determinism but instead for abnormality and potentiality. However, the idea of politics as the direct expression of life's potentiality is affected by the same short-circuitry as that on which the biopolitics of the twentieth century was ultimately based: the apparently emancipatory invocation of a vital imperative fashionably dressed up in the terms of biological plasticity inevitably ends up serving as an unquestionable normative ground for politics. The free, creative, and nomadic essence of life donates to the living being its inessential essence, and this gift prescribes a duty, namely, an infinite demand for liberation and contamination. The only possible politics will be a biopolitics able to give voice to the erratic essence of life itself, in the form of a 'plural and potentially rebellious multitude' (Esposito 2006, 165); 'ironically, this vision of a purely immanent self-organizing and self-governing community of men may be read as a *manifesto* of biopower rather than an articulation of resistance to it' (Prozorov 2007, 64), which implies 'unconditional valorization of the plenitude of biopoliticized existence as an emerging force of freedom' (Prozorov 2007, 64). The vitalist approach entails that, if biology has intended human beings to be flexible, those who do not conform to this flexibility are inevitably against nature and therefore enemies of the survival and reproduction of biological life itself.

But in order to render biopolitics compatible with the modern theory of the state shouldn't one embrace a purposeless idea of life? Should one, in other words, find in the void of meaning resulting from the awareness of the lack of any political design intrinsic to life itself not a nihilist conclusion to be filled with new transcendent meanings, but the premises for an affirmative and democratic biopolitical rationality? And isn't the attempt at finding a more originary ground for politics in life itself rather than simply in the 'artificial person' of the state a theoretical move contrary to the founding gesture of modern political theory?

THE DANCING GOD

The vitalist approach defines a conception of the human being which is clearly incompatible with the one presupposed by the discursive one. The mutual exclusion between these two notions of the human being does not conclude the discussion on them, since this mutual exclusion becomes the way in which these two definitions can be together and belong to each other.

In the previously mentioned passage where Foucault names Heidegger as one of the most relevant figures for the development of his work, he also reveals that it is Nietzsche who is his main guiding figure: 'my entire philosophical development was determined by my reading of Heidegger. But I recognize that Nietzsche prevailed over him' (Foucault 1988, 250). Foucault's predilection for Nietzsche over Heidegger is not intended as a preference for the vitalist over the discursive approach. The influence of Nietzsche is far more radical than that, since it is only thanks to Nietzsche that it is possible to name that ontological horizon shaped by the conjunctive antithesis between the 'impatience for liberty' and the impatience for life. In *Thus Spoke Zarathustra*, Nietzsche writes: 'I would only believe in a god who knew how to dance' (Nietzsche 2006, 29). I believe that the most appropriate name for the secret link between the 'impatience for liberty' and the impatience for life has to be that of Dancing God, since the idea of the dance expresses the idea of movement and constant transformation embedded in the Foucauldian conception of impatience. I maintain that the Dancing God is the name of that limbo in which the process of deconstruction of the ultramundane origin of the human being summed up in the incisive statement about the death of God has stopped. The 'impatience for liberty' and the impatience for life are the two ways in which the legacy of this unfinished process of secularisation is interpreted by Foucault. If the notion of Dancing God cannot be fully developed here, it is worth noting that thinking beyond the Dancing God means to think about the human being and his origin without embracing any anthropomorphic and teleological principle. The abyssal thought of a human being who is not *for* anything subtracts him also from the destiny of any dance and draws the topology of a different land. But this new ontological terrain where a different biopolitical rationality can grow requires a redefinition of the relationship between naturalism and critical theory.

NOTES

1. The term is borrowed from Roberto Esposito (2006, 175).
2. Alenka Zupančič proposes a convincing analysis of Foucault's 'repressive hypothesis' in "Biopolitics, Sexuality and the Unconscious," *Paragraph* 39 (1), 49–64. Boštjan Nedoh addresses the legacy of Foucault's 'repressive hypothesis' in current Italian biopolitical theory in "Alive or Undead? Biopolitics between Esposito's Vitalism and Lacanian Psychoanalysis," *Paragraph* 39 (1): 65–81.
3. Despite Butler's thought being very articulate and subject to significant developments in the course of the years, the perspective from which she issues her critique to Foucault in this 1989 article is a very good example of the discursive approach.
4. I have addressed this idea in more detail elsewhere (see Piasentier 2016). Here I limit myself to exposing the key points of this account of life, to then outline the

implications it raises for the understanding of the Foucauldian vitalist approach to biopolitics.

5. For a thoughtful and clear critique of the current concept of life, see Tarizzo 2011 and 2017.

BIBLIOGRAPHY

Agamben, Giorgio. 1998. *Homo Sacer: Sovereign Power and Bare Life*. Palo Alto, CA: Stanford University Press.

Agamben, Giorgio. 2004. *The Open*. Palo Alto, CA: Stanford University Press.

Agamben, Giorgio. 2016. *The Use of Bodies*. Palo Alto, CA: Stanford University Press.

Aristotle. 2002. *De Anima. Books II and III (with Passages from Book I)*, translated by D. W. Hamlyn. Oxford: Oxford University Press, Clarendon Aristotle Series.

Ayala, Francisco J., and Robert Arp. 2010. "Have Traits Evolved to Function the Way they Do Because of a Past Advantage? Introduction." In *Contemporary Debates in Philosophy of Biology*, edited by Francisco J. Ayala and Robert Arp, 49–52. Oxford: Wiley-Blackwell.

Braver, Lee. 2007. *A Thing of This World: A History of Continental Anti-Realism*. Evanston, IL: Northwestern University Press.

Butler, Judith. 1989. "Foucault and the Paradox of Bodily Inscriptions." *Journal of Philosophy* 86 (11): 601–7.

Chomsky, Noam, and Michel Foucault. 2006. *Human Nature: Justice versus Power: The Chomsky–Foucault Debate*, edited by Fons Elders. London: Souvenir Press.

Cummins, Robert, and Martin Roth. 2009. "Traits Have Not Evolved to Function the Way They Do Because of a Past Advantage." In *Contemporary Debates in Philosophy of Biology*, edited by Francisco J. Ayala and Robert Arp, 72–86. Oxford: Blackwell.

Darwin, Charles. 2002. *Autobiographies*, edited by Michael Neve and Sharon Messenger. London: Penguin.

Dreyfus, Hubert, and Paul Rabinow. 1983. *Michel Foucault: Beyond Structuralism and Hermeneutics*. Chicago: University of Chicago Press.

Esposito, Roberto. 2006. *Bíos: Biopolitics and Philosophy*. Minneapolis: University of Minnesota Press.

Esposito, Roberto. 2011. "Politics and Human Nature." *Angelaki: Journal of the Theoretical Humanities* 16 (3): 77–84.

Foucault, Michel. 1972. *The Archeology of Knowledge and the Discourse on Language*. New York: Pantheon Books.

Foucault, Michel. 1977. *Discipline and Punish: The Birth of the Prison*. New York: Pantheon Books.

Foucault, Michel. 1978. *History of Sexuality*, vol. 1. New York: Vintage Books.

Foucault, Michel. 1988. "The Return of Morality." Interview conducted by Gilles Barbedette and André Scala, in *Politics, Philosophy, and Culture: Interviews and Other Writings, 1977–1984*, edited by L. Kritzman, 242–55. New York: Routledge.

Foucault, Michel. 1994. *Essential Works of Foucault 1954–1984*, vol. 2: *Aesthetics, Method, and Epistemology*, edited by James Faubion. London: Penguin Books.
Foucault, Michel. 2000a. *Ethics: Essential Works of Foucault 1954–1984*, vol. 1, edited by Paul Rabinow. New York: Penguin.
Foucault, Michel. 2000b. *Power: Essential Works of Foucault 1954–1984*, vol. 3, edited by James D. Faubion. New York: Penguin.
Foucault, Michel. 2005. *The Order of Things: An Archeology of Human Sciences*. London: Routledge.
Foucault, Michel. 2007. *Security, Territory, Population: Lectures at the Collège de France 1975–1976*. New York: Picador.
Gadamer, Hans-Georg. 1981. *Reason in the Age of Science*. Cambridge, MA: MIT Press.
Gould, Stephen J. 1980. *The Panda's Thumb*. New York: W. W. Norton.
Gould, Stephen J. 2002. *The Structure of Evolutionary Theory*. Cambridge, MA: Belknap Press of Harvard University Press.
Heidegger, Martin. 1998. "Letter on 'Humanism.'" In *Pathmarks*. Cambridge: Cambridge University Press.
Lemke, Thomas. 2011. *Biopolitics: An Advanced Introduction*. New York: NYU Press.
Malabou, Catherine. 2008. *What Should We Do with Our Brain?* New York: Fordham University Press.
Nedoh, Boštjan. 2016. "Alive or Undead? Biopolitics between Esposito's Vitalism and Lacanian Psychoanalysis." *Paragraph* 39 (1): 65–81.
Nietzsche, Friedrich W. 1996. *Human, All too Human*. Cambridge: Cambridge University Press.
Nietzsche, Friedrich W. 2006. *Thus Spoke Zarathustra*, edited by Robert Pippin. Cambridge: Cambridge University Press.
Nietzsche, Friedrich W. 2011. *On the Genealogy of Morality*. London: Hackett.
Piasentier, Marco. 2016. "The Vital Error: Where Evolutionary Biology and Genealogy Meet." *Paragraph* 39 (1): 93–107.
Prozorov, Sergei. 2007. "The Unrequited Love of Power: Biopolitical Investment and the Refusal of Care." *Foucault Studies* 4 (February): 53–77.
Tarizzo, Davide. 2011. "The Untamed Ontology." *Angelaki, Journal of the Theoretical Humanities* 16 (3): 54–55.
Tarizzo, Davide. 2017. *Life. A Modern Invention*. Minneapolis: University of Minnesota Press.
Zupančič, Alenka. 2016. "Biopolitics, Sexuality and the Unconscious." *Paragraph* 39 (1): 49–64.

Chapter 2

The Life Function

The Biopolitics of Sexuality and Race Revisited

Jemima Repo

Biopolitics is analysed increasingly through the deadly operation of racialisation. Race is a discourse through which the human species is divided into groups and accordingly regulated, normalised and administered (Stoler 1995, 9). In his *Society Must Be Defended* lectures, Foucault spoke of the 'death function' (Foucault 2003b, 258) or 'death instinct' (1981, 156) as 'the murderous function of the State' (2003b, 256). With this he did not simply mean crude killing but forms of 'indirect murder' (2003b, 256), such as the exposure of someone to a greater risk of death, political death, expulsion, rejection, discrimination and so on. Foucault located the concentration of the death function in modern biopolitics as bound to the discourse of racism, whilst the discourse of life operated through the discourse of sexuality. In this chapter,[1] however, I argue that neither the functions[2] of death nor life are historically stable in their respective attachments to race and sexuality, and therefore caution should be exercised when ascribing all deadly biopolitical practices to racialisation. As the chapter demonstrates, both homosexuality and heterosexuality have been apparatuses of death and life, suggesting that the life and death functions of biopolitics manifest themselves through historically varying and specific discourses of difference.

Building on Foucault's brief analyses of race in *Will to Knowledge* and the *Society Must Be Defended* lectures, thinkers such as Agamben (1998), Esposito (2008) and Mbembe (2003) have developed theorisations of the 'thanatopolitical' or 'necropolitical' aspects of biopolitics. While the emphasis on race and death in such Foucauldian scholarship is timely and welcome, it often overlooks the relevance and relationships they have to the biopolitics of sexuality and life. Biopolitics itself becomes theorised through death function with three significant consequences. First, it prioritises the murderous effects of biopolitics as the primary active forces that penetrate and administer life. Second, it therefore reduces biopolitical difference production to racialisation and death; and third, it does so at the expense of downplaying the forceful endeavour to reproduce life. Finally, Jin Haritaworn (2012, 13) and Alexander Weheliye (2014, 38) have argued, by adhering to an overly general race that collapses all modes of difference into a single concept of race, we run the risk of assuming that all differentiations are produced and regulated through a single logic of racialisation. Further inspection of the genealogies of sexuality and race is necessary in order to understand how they are differently deployed as biopolitical apparatuses, with a focus on their historically contingent strategic underpinnings of life and death.

In response to this dilemma, this chapter introduces the concept of the 'life function' as a way of mapping the biopolitical facilitation of the reproduction of population and thus life. I employ it to revisit the biopolitical strategy of fostering life by 'distributing the living' (Foucault 1981, 144) through the apparatus of sexuality. The chapter re-examines not only the genealogies of sexuality and race but also the tactical and strategic facilitation and suppression of life in relation to these apparatuses. The result is a more nuanced understanding of not only how the racial production of difference varies from the sexual production of difference but also of the contingent and often complex relation between the rationalities of fostering life and forbidding it and the apparatuses deployed to carry them out. We can witness, for example, how the death function can also operate through sexuality, and how the discourses of heterosexuality and homosexuality have historically shifted from the realm of death to life.

Following a re-examination of the *Society Must Be Defended* lectures and *Will to Knowledge*, this chapter maps out Foucault's account of the tactics of difference production through a critical discussion of the biopolitical strategies by which sexuality and race are produced. I then discuss how sexuality has also been an apparatus for administering the death function. Finally, through a discussion of the shifting biopolitical strategies underpinning 'homosexuality' and 'heterosexuality', I examine how sexuality has been invested historically with contingent imperatives of life and death without being reducible to racialisation, but always related to it. Specifically, I examine how heterosexuality was once consigned to the realm of deviance and death, and likewise how homosexuality is currently being recaptured and reorganised, or 'reterritorialized' (Deleuze and Guattari 2004, 434), by the life function in current debates of gay marriage and parenting. Through these examples, I argue that the functions of life and death are not exclusively tied to the apparatuses of sexuality and race, respectively, but rather these apparatuses are deployed in different historical moments and contexts to perform the administration of the life of the species, often involving radical reversals of discourse, strategy and tactics.

FOUCAULT'S GENEALOGIES OF RACE AND SEXUALITY

Before revisiting Foucault's genealogy of race, it should be pointed out that there are two versions of them available, and both of them are in their own ways tentative and incomplete. Prior to the publication of the *Society Must Be Defended* lectures, Foucault's writings on race were limited to a few pages in the first volume of *The History of Sexuality*. These two analyses of race, as Ann Laura Stoler (1995) explains, have significant differences. In the lectures, Foucault's analysis of racism is primarily linked to biopower and the emergence of nationalism, whereas in *The Will to Knowledge*, it is bound to sexuality and the bourgeois order. Racism takes a central position in the lectures, only to be repositioned, according to Stoler, as an effect of the discipline of sexuality in his published work. For Stoler, not only did Foucault ignore colonialism in his genealogies, but also the shift of his analytical focus from race to sexuality entailed 'a clean erasure of the question of racism from his project' (1995, 25). Race certainly features less prominently in *Will to Knowledge*, but it is not erased or subordinated to sexuality in it. Rather, sexuality is identified as the critical hinge through which race was biopoliticised. Whereas in the lectures race and sexuality stood as separate apparatuses of biopower, *Will to Knowledge* provides a slightly different thesis about the genealogical entanglement and interdependence between race and sexuality.

In the *Society Must Be Defended* lectures, race does not appear initially as a biologically founded racist or repressive discourse but as the discourse of a race struggle or race war used to the attack of the unjust possession of power by another group (Foucault 2003b, 254). According to Foucault, in the nineteenth century race became the discourse that legitimised the state as the protector and administrator of the affairs of the race. This, for Foucault, marked the emergence of state racism and the flourishing of the biological discourse of race in the mid-nineteenth century.[3] Race was no longer a discourse of the oppressed concerned with emancipation but one oriented to delineate and sustain the racial purity of a population. The preceding context of struggle as battle was replaced by an evolutionary theme of struggle, where the struggle was one of natural selection for the survival of the species (Foucault 2003b, 254). This discourse of biological race struggle provided the state with a justificatory framework for its murderous function 'by appealing to the principle that the death of others makes one biologically stronger insofar as one is a member of a race or a population' (Foucault 2003b, 256). Thus the discourse of state racism enabled the biopolitical state to kill not just those of other states but also its own citizens.

In the lectures, Foucault states somewhat vaguely that sexuality was a 'very different'—although 'not altogether that different' (2003b, 256)—axis of biopolitics from race. In the last lecture, Foucault recognises the 'privileged position' (2003b, 252) that sexuality assumed in the nineteenth century. He already notes the different forms, functions and modes of their emergence in the biopolitical. 'The emergence of biopower', according to Foucault is what 'inscribes [racism] in the mechanisms of the state' (2003b, 256). Sexuality on the other hand, 'was important for a whole host of reasons'. First, sexuality was related to 'corporeal mode[s] of behaviour'. Second, it 'also has procreative effects' through the inscription of biological bodily processes. Finally, sexuality 'occupies a privileged position between organism and population, between the body and general phenomena' as it is where discipline (that acts on the body) and regulation (the biopolitics of population) intersect. In the lectures, race and sexuality therefore seem to circulate as two separate strands of biopolitical intelligibility, where the former is divisive and deadly, while the latter is corporeal, behavioural and procreative.

By contrast, in *Will to Knowledge* we are treated to a genealogy of sexuality that prioritises the question of 'life' and 'its unfolding', where both sexuality and race are mobilised under the politics of fostering life. Racial discourse differentiated between species, debating whose lives should be promoted and regulated, while sexuality posed the question of the normalisation of bodies and reproduction of life. Specifically, Foucault articulates sexuality as the hinge between the 'anatamo-politics of the human body' and the 'biopolitics of the population' (Foucault 1981, 139), which rendered sexuality 'a mean so access to both the life of the body and the life of the species' (1981, 145–46). It enabled the control of both the body and the population as a whole, relying on a whole host of mechanisms (e.g., medical, biological, pedagogical, demographic and psychiatric) for the 'state management of marriages, births, and life expectancies' (1981, 118). In other words, sexuality was recognised as a means to control the reproduction of the fleshy, material object of biopolitics, the species-body. To grasp and maximise control over this function, sexuality was deployed through the motif of the Malthusian couple, the socialisation of procreative behaviour, as well as the hysterisisation of women's bodies and the pedagogisation of children's sex (1981, 104–105). Sexuality thus was not just a means of inserting the human body into the capitalist machine and ensuring labour capacity. It ensured the maintenance of population numbers, particular bodily behaviours and re/productive capacities. In the *Will to Knowledge*, Foucault writes that state racism took shape in the second half of the twentieth century—the same time as the sexual apparatus—and that 'it was accompanied by a long series of permanent interventions at the level of the body, conduct, health, and everyday life', and was driven by the need to 'ensur[e] the triumph of the race' (1981, 149). Sexuality is the point of access to both the population and the bodies that constitute it, but without race there would be no defined or delineated 'living' species population to target its disciplinary and regulatory controls. The result of this convergence was the development of the eugenic ordering of society that bridged race and sexuality as two mutually supporting apparatuses of modern biopower.

It is for this reason that it is difficult, if not problematic, to pit 'biopolitics' against 'thanatopolitics' or 'necropolitics' as separate domains or modes of biopower. Foucault, after all, reminds us that the death function of race does not take killing as its end goal, but rather it is the maintenance and administration of the life of the species. As Robert Bernasconi (2010, 209) observes, 'when sex became a police matter, so did race, but not merely in the sense that both were regulated, but because . . . it was through the controlling of sexuality that the potential of a race is maximized'. As the concept of intersectionality (Collins and Bilge 2016) already insists, also in biopolitics discourses and race and sexuality should therefore not be treated as separate or oppositional but as necessarily complementary. For there is no 'species' to speak of or defend without discourses of race, and the species cannot survive or flourish without its reproduction.

SEXUALITY AND DEATH

Recent feminist scholarship has done much to analyse the complex relationship between racial and sexual biopolitics (e.g., Apostolidis 2011; Deutscher 2012; Haritaworn 2012; Lettow 2015; Repo 2016; Schultz 2015; Stoler 1995), advancing the argument of the interdependence of these two biopolitical apparatuses. Lettow (2015) for instance, has shown how the concept of reproduction was constitutive of the formation of the genealogical concept of race. Moreover, she argues that reproduction emerged as a nodal point for three biopolitical problems: the management of populations, the politics of racialisation, and the reorganisation of kinships and sexed relations (2015, 268). Indeed, it is the focus on reproduction that has long brought into focus the intersecting biopolitics of sexuality and race in feminist thought, even though it has not always been articulated in these terms. Anna Marie Smith (1994), for example, has analysed British immigration legislation that placed more restrictions on British women than British men who brought foreign spouses into the United Kingdom. The 1914 British Nationality and Status of Aliens Act ruled that only foreign women, not men, who married British men were eligible for British nationality. The law was underpinned by a fear that presumably nonwhite male immigrants would become sexually involved with assumedly white British women if they were forbidden to bring their wives and children into the country. In the eyes of the British government, nonwhite female immigrants lacked the hypersexual predation of the black male, seen as a threat to the racial purity of the white British population. The act therefore sought to protect the racial purity of the British nation by guarding white British reproductive femininity against a racialised and pathologically sexualised masculine other. Smith's work therefore exemplifies how the life function and death function necessarily operate in unison: racialisation occurs as an inextricable part of the strategy of the sexual reproduction and regulation of the life of the species.

While such examples add complexity to our understanding of how the biopolitics of death and life intersect through race and sex/sexuality respectively, the death function also unfolds through sexuality. The relation between functions and the apparatuses through which they are realised is, I argue, a contingent one. Gilles Deleuze and Félix Guattari's (2004, 434; 442) concept of re/deterritorialisation is useful to understand this relation. Deterritorialisation and reterritorialisation refer to interlinked processes that are in constant flux, the former involving something being left behind or breaking apart, and the latter involving the recombination of elements leading to the emergence of new assemblages and relations, such as new apparatuses. Through this conceptual pair, it is also possible to consider the historically and contextually contingent shaping and reshaping of apparatuses of power and how they become anchored to and from the functions of life and death. The biopolitical production of race, for example, is the effect of the reterritorialisation of certain biological, psychological and sociological knowledge by the death function, categorising the human species into various groups and subgroups. Likewise, the biopolitical production of sexuality is an effect of the reterritorialisation of the life function, premised upon, for example, linkages forged between biological knowledge of reproductive organs and the psychiatrisation of pleasure. In this section I employ the concepts of re/deterritorialisation to argue that the death function is not exclusive to race. The death function can become reterritorialised; in other words, its elements and relations can be recombined and recaptured anew, around and through sexuality.

Homophobia is perhaps the most obvious example of the death function operating through sexuality. The deadly and exclusionary mechanisms enacted through sexuality are strategically distinct from the death function of race. In this case, the death function of sexuality disciplines the corporeal behaviour of the sexed and sexual subject—in other words, attempts to subject it to the life function by urging, persuading or coercing it to sexually reproduce at the threat of death. The territorialisation of sexuality by the death function re-enforces the discipline of the category of sexuality established by the life function. In other words, death can be reterritorialised into sexuality as the ultimate punishment for a sexuality that will not discipline and normalise itself to perform this function.

The infamous murder of Matthew Shepard is a case in point. On the evening of the 6th of October 1998, twenty-one-year-old homosexual US university student Matthew Shepard was kidnapped by two other twenty-one-year-old young men in Colorado. Shepard was brutally beaten with a pistol, resulting in severe bruising and crushing parts of his skull, then stripped naked and tied to a fence and abandoned. He was discovered the next afternoon, taken to hospital, and died a few days later on the 12th of October. While the case has been covered by several feminist and queer theorists in relation to gender violence (Butler 2004; Loffreda 2000; McWhorter 2009; Swigonski, Mama and Ward 2001), it can also be considered to be an example of how the death function is reterritorialised through the motif of sexuality. Shepard's intolerable homosexuality became punishable by death. Death, in this instance, was brought to bear on his person as the final form of discipline for sexuality that did not behave itself. The Shepard case therefore suggests that for life to be allowed to die does not necessarily require the death function of race as such. Whereas racism, in Foucault's analysis, excludes on the basis of assumed species difference, in this case homophobia excludes according to the perceived misbehaviour of a sexed body and its aberrant sexual desire in relation to other sexed bodies. Homophobia can in such instances be understood as 'an oppression based on the *activities* of the members of a group, and not any definite group attributes' (Grosz 1995, 225), like racism most often is. Sexuality becomes reterritorialised by the death function when the perceived sexual misbehaviour becomes not only intolerable but threatening, and those practicing it must be left to suffer, die or be killed.

While the death function therefore operates at its most intense through race, nothing dictates that it should be exclusive to it. Thus, the prioritisation of the enmeshment of the murderous state with the apparatus of race in post-Foucauldian approaches to biopolitics is complicated, for example, by the fact that those bearing the pink triangle were not dispatched to the camps because they were not of the Aryan race (though some of them surely were both homosexuals and non-Aryans), but because their sexual deviancy was seen as a threat to the reproductivity of the Aryan race. Homosexuals were not confined and killed because they were racially different but because their sexual behaviour threatened the reproductive capacities of the Aryan race. This distinction is essential, as it argues that when the death function territorialises sexuality, it does so through a different tactical trajectory than race.

Foucault tells us that sexuality is bound to the production of life, not death, which may explain why so few feminists have taken up the problem of biopolitical death in relation to sexuality/sex.[4] One exception is Judith Butler, who has argued that in the 1980s AIDS crisis we can witness how '"sex" is constructed not only in the service of life or reproduction but . . . in the service of the regulation and apportionment of death' (1996, 61). According to Butler (1996, 72), Foucault was too preoccupied with disentangling the psychoanalytic association of sex with death that he overlooked how sex not only secures life but produces and proliferates death. Nonetheless, she acknowledges that Foucault was aware that '"sex is worth dying for" . . . that preserving the regime of "sex" is worth dying for and that political wars are waged so that populations and their reproduction can be secured' (Butler 1996, 73). Indeed, Foucault (2003a, 60–61) made clear that sexuality is not just exclusively a question of reproduction and vitality, but because it is also a fundamentally medicalised discourse, it operates on the basis of normalisation to the extent of rendering monstrous those that it consigns to the abnormal.

In turn Susan Sontag argues that in the 1980s US public discourse, AIDS was seen as 'a punishment for deviant sex' (1990, 151), which is why many felt that those infected could be 'deservedly' left to die. The homosexual subjects of 1980s AIDS discourse were produced and condemned not only by the sexual apparatus but also a racial one. It was not just a case of reprimanding or allowing punishment to fall on sexual deviants but also disciplining a racial and eugenic threat. Its carriers were ostracised for allegedly participating in sexual (mis)conduct that spread an infectious disease that was detrimental to the health and potency of Western populations. Moreover, in the West, the African origin of the HIV virus played a significant role in framing AIDS a major racial threat; it is consistently portrayed as 'a disease that comes from foreigners, from outsiders, and especially from "black" Africans' (Elbe 2005, 411). The death functions of race and sexuality were deployed to safeguard the life function of Western populations, which needed to be protected from a debilitating disease from a racially inferior African continent. The effort to defend the population from the corrupting malady is exemplified by immigration policies of the United States (and also some non-Western countries such as China), in force until 2010, that restricted the mobility of persons diagnosed with HIV/AIDS, barring them from entering the country, for example, by refusing to issue visas to HIV-positive applicants.

In this section I have shown that the death function is not exclusive to race, but that it can also territorialise sexuality. Whereas race becomes reterritorialised by the death function as a means of establishing the parameters of the species, of those who should be allowed to live and encouraged to reproduce and who should not, the death function is reterritorialised in sexuality when a perceived sexual misbehaviour becomes intolerable or threatening, and those practicing it must be left to die or be killed in order to protect the propagation and well-being of the species. In the human species, therefore, the death function operates through both race and sexuality, whereas the life function is territorialised on sexuality, and specifically heterosexuality. However, with further analysis, even the life function's straightforward territorialisation of heterosexuality cannot be taken for granted.

HETEROSEXUALITY: FROM DEATH TO LIFE

Having explored the relationship between race and sexuality, and between the life and death functions, this section returns to the concept of the life function. I have conceptualised the life function as the endeavour to reproduce life, and thus, the reproduction of the object of biopolitics. At present, heterosexuality dominates as the discourse through which the life function is most rigorously deployed. It cannot, however, be taken for granted as the essential procreative, life-administering discourse. In fact, Jonathan Ned Katz's (2007) *The Invention of Heterosexuality* demonstrates that heterosexuality did not properly enter the realm of normality until the early twentieth century. In a provocative genealogy of heterosexuality, Katz demonstrates that 'heterosexuality is not identical to the reproductive intercourse of the sexes' nor is it 'the same as sex distinctions and gender differences', or even 'the eroticism of women and men'. Rather, it is 'one particular historical arrangement of the sexes and their pleasures' (Katz 2007, 14). Katz does not contradict Foucault's grounding of the deployment of sexuality in the Victorian era but rather expands his genealogy. After all, in *Will to Knowledge* Foucault (1981, 3; 105) did not analyse a discourse of heterosexuality but rather identified the Malthusian couple in the conjugal bourgeois family as the figures problematised to realise procreative reproduction (Deutscher 2012; Repo 2014). Katz's genealogy serves as a warning not to conflate this with heterosexuality, as the Mathusian couple of the conjugal family is merely one form or ordering of different-sex sexual relations that is not equivalent to heterosexuality.

Indeed, Katz's genealogy demonstrates that heterosexuality was not initially territorialised by the life function. As Katz writes, 'the new term *heterosexual* did not, at first, always signify the normal and good' (2007, 19). In the late Victorian period the term was first employed by psychiatrists who equated it with perversion. In 1892, Dr James G. Kiernan of Chicago, for example, wrote in an article on 'Sexual Perversion' that heterosexuality was one of the 'abnormal manifestations of the sexual appetite' (Katz 2007, 20). Such definitions circulated until the 1920s, when an edition of Webster's dictionary still defined heterosexuality as a 'morbid sexual passion for one of the opposite sex' (Katz 2007, 92). Heterosexuality deviated from the norm of sex as the dominion of true love that could only be fulfilled after marriage. Well into the early twentieth century, therefore, heterosexuality was an aberration that referred to an unproductive different-sex pleasure that separated lust from love.

On the basis of Katz's genealogy, the normality of all different-sex erotic relations cannot be taken for granted. The 'heterosexual' is not a self-evident subject with a coherent history but rather has been shaped by a complex process of de-/reterritorialisations around different-sex sexuality. Heterosexuality was gradually normalised from the turn of the century onwards, at first by individual psychiatrists claiming that the sexual instinct of heterosexuality was biological and therefore natural. Later they co-opted the term 'to affirm the superiority of different-sex eroticism' (Katz 2007, 52–53). Katz explains the change in terms of the increasingly common private, nonprocreative practices of erotic pleasure amongst the late nineteenth-century middle classes that contradicted the norm of conjugal sex. By then, the middle classes had established their social position and importance and the expanding prevalence of 'deviant' heterosexual behaviour began to menace the dominance of the figure of the Malthusian couple through which the life function operated. Consequently, the threat posed by different-sex pleasure was defused through the re-invention of heterosexuality as a discourse that 'publicly named, scientifically normalized, and ethically justified the middle-class practice of different-sex pleasure' (Katz 2007, 51). Thus, nonprocreative different-sex sexuality came to be established as a norm rather than a deviation.

Katz (2007, 58) proceeds to examine Freudian psychoanalysis as a new field of knowledge that both supported and challenged heterosexuality. On the one hand, Freud problematised homosexuality as a developmental aberration and assumed heterosexuality as the successful result of early development. On the other hand, Freud denaturalised heterosexuality by rendering it a social product of the development process. Rephrased in Foucauldian terms, making heterosexuality psychiatrically and socially acceptable signified not a dismantling of the sexual apparatus but an expansion of its terrain through a shift in its configuration.

Freud's work accomplished the legitimation of the previously 'morbid' heterosexuality through at least two significant shifts in the understanding of sexuality: first, by adding pleasure to the elements that are seen to constitute life, and second, by laying new importance to the social, nonreproductive aspects that ensure the reproduction of different-sex sexual relations. Freud radically rearranged the sexual order of things to argue that pleasure, 'the attainment of organ-pleasure' (Freud 2001, 25), not reproduction, is the main driver of the sexual instincts. In this new schema, reproduction is subordinated to the pleasure principle but not entirely replaced and certainly not erased by it. Pleasure became the main goal of the sexual instinct, and fertility was rendered 'a late, secondary development in life's long pursuit of happiness' (Katz 2007, 60). In Freud, we could say, pleasure is reterritorialised by the life function through a shift in the sexual order of things away from the crude reproductive principle to a more complex apparatus that now incorporates pleasure.

Freud toppled the dominance of reproductive imperative, but in so doing he rendered pleasure essential for reproduction, for given the primacy of pleasure, there could no longer be sexual reproduction without the primary initiating drive for pleasure. For Freud, the aim of each of the sexual instincts, which 'emanate from a great variety of organic sources' (Freud 2001, 124) is to aim for the 'attainment of organ-pleasure' upon which 'do they enter the service of the reproductive function and thereupon become generally recognized as sexual instincts' (2001, 125). Pleasure is therefore not a guarantee for reproduction, but reproduction arises from the pleasure-seeking instinct of the individual. A sexual instinct is defined by the eventual entering of an instinct into the service of the reproductive function. An instinct can be labelled as perverse when the acts move completely away from reproduction and exclusively carry out a given deviation (Katz 2007, 60). A normal sexual instinct for Freud can therefore accommodate a whole host of practices before it becomes deviant, and this deviancy is ultimately defined by its removal from the reproductive. The introduction of the pleasure principle therefore expanded the kinds of relations that could be infused with biopower as potential vessels of the life function. Different-sex pleasure-seeking eroticism therefore went from being corruptive of the life function to conductive of it. By linking pleasure to life, the previously sexually deviant heterosexuality became a potentially biopolitically useful discourse of subjectivity.

The Freudian discourse also introduced the notion that the nuclear family's economy of desire was responsible for heterosexual and therefore procreative socialisation. As Guy Hocquenghem argued, 'Freud discover[ed] the libido to be the basis of affective life and immediately enchain[ed] it as the Oedipal privatization of the family' (1993, 73). The linking of pleasure to life occurred hand in hand with the emergence of the Oedipal thesis. The second tactical shift apparent in Freud's work is the socialisation of heterosexuality through the Oedipal process, which purports that a boy child becomes heterosexually socialised by wanting to sexually possess his mother and kill his father. (Conversely a girl sexually desires her father as a consequence to the experience of penis envy.) As Katz argues, there is a subversive aspect to this theory in that it argues that 'heterosexuals are made not born' (2007, 74). Yet, at the same time, the notion of the Oedipus complex renders the successful attainment of heterosexuality a biopolitical challenge. The emergence of the Oedipal scenario meant that different-sex sexual attraction could no longer be taken for granted. Rather, it had to be ensured through the appropriate upbringing of the child. Thus, the sexuality of both the child and parent were problematised, reorienting and expanding the reach of biopower to a new set of relations and relationships through which the life function could operate.

The notion that the child must desire the parent of the opposite sex also relates back to how pleasure became bound to sex and life. The child must learn to desire 'correctly' by yearning for the parent of the opposite sex, and the parent's response to the child's interest would determine the child's future orientation for pleasure-satisfying objects and sexes. Pleasure, in this sense, was essential for educating the sexual instincts in such a way that they would eventually lead to reproduction.

The challenge posed to the Malthusian conjugal couple by heterosexuality therefore resulted not in the liberation of sexuality but the reterritorialisation of the life function on heterosexuality. It did not replace the Malthusian couple but rather expanded the range of different-sex sexualities territorialised by the life function. Heterosexuality, originally ascribed to abnormality and death, re-emerged as a vehicle for the reproduction of life. Since the 1970s, however, heterosexuality too has been challenged. The rise of divorce rates, the articulation of new family forms, and the feminist and gay liberation movements have all been perceived as destabilising forces for the conjugal and/or heterosexual apparatus. According to Katz, 'the commodification of pleasure further breaks down old distinctions between hetero and homo' by marketing pleasure-sex to both sexualities, whereby 'heterosexuals are more and more like homosexuals, except for the sex of their sexual partners' (2007, 187). Others have commented on the selectivity of this new 'homonormativity' (Duggan 2003) that heteronormalises homosexuality. Smith, for example, demonstrates how in the 1980s in Britain the acceptable homosexual was a 'law-abiding, disease-free, self-closeting homosexual figure who knew her or his proper place on the secret fringes of mainstream society' (Smith 1994, 18). Positive-image campaigns promoted an image of nonthreatening 'white, middle-class professional gay men in monogamous relationships who demonstrate impeccable bourgeois taste' (1994, 190). This normalisation, I argue, also entails a more recent reterritorialisation of homosexuality by the life function.

THE RE-OEDIPALISATION OF HOMOSEXUALITY

If a renewed biopoliticisation of homosexuality is not immediately apparent from the marketing techniques mentioned above, it becomes so after a glance at recent Western scientific and political debates on monogamous same-sex relationships. Notably, the struggle for the political recognition of gay relationships through the institution of marriage is often paired with a debate about same-sex parenting. In the Netherlands in 2001 and in Norway and Sweden in 2009, adoption rights were also passed at the same time as gender-neutral marriage laws replaced registered partnerships previously available to same-sex couples. In Spain in 2005, adoption rights were passed with the same stroke as same-sex marriage was legalised. Marriage continues to be so bound to the norms of family and parenthood that to approve of same-sex marriage seems to be equated with the approval of same-sex parenting. Indeed, it is in this latter point that prominent voices from the scientific community concentrate their efforts.

Psychiatric research in particular is used by both proponents and opponents of same-sex marriage to argue that same-sex parenting either does not or does harm the sexual development of children (Clarke 2000, 155). When US professor of law Lynn D. Wardle (1997, 846–47) initiated a debate on homosexual parenting in 1997 in the *University of Illinois Law Review*, he argued that different-sex parents are necessary for the balanced psychological development of a child. Fatherlessness, for example, can lead to friendlessness, poorer educational standards and lack of discipline, crime, emotional insecurity and unhappy and unsuccessful future relationships. Wardle's article leaned on a common conservative argument against same-sex parenting that claimed that homosexuals are unable to raise normal—that is, heterosexual—children (e.g., Hicks 2005, 163). Referencing psychoanalytic thinkers like Erik Erikson and Bronislaw Malinowski, he relied on the Freudian Oedipal scenario to explain the 'correct' development of children's sexuality. In other words, for Wardle, homosexuality threatens the Oedipal process (and thus the reproduction of the species). Not only does it not reproduce society, but it also does not reproduce the Oedipus complex necessary for its reproduction, which is ensured through the heterosexual family (Hocquenghem 1993, 106).

Defenders of gay marriage and parenting, by contrast, have counterargued that 'lesbian and gay parents do not produce inferior or even particularly different kinds of children than do other parents' (Stacey 1996, 129). They do not question the overall validity of the Oedipal discourse that addresses the relationship between child and parent but rather argue within the finer points of the apparatus of sexuality. In so doing they too engender the reproblematisation of homosexuality that also amounts to a re-Oedipalisation of homosexuality through new tactical shifts that modify the Freudian Oedipal thesis to accommodate gay parenting within the sphere of normality. A number of scholars since the 1980s, such as William Meezan and Jonathan Rauch, refute the traditional Freudian reading according to which 'children of lesbian and gay parents are confused about their gender identity . . . or that they are more likely to be homosexual' (Meezan and Rauch 2005, 103). Instead, they suggest that children might actually 'benefit from the marriage of their lesbian and gay parents' (2005, 107). First, they write, 'marriage increases the economic capital available to children'; second, the 'durability and stability of the parental relationship . . . is of vital importance to children' (2005, 108); and third, 'same-sex marriage might benefit children through social investment' by providing them with 'social resources', 'legal and regulatory protections' and intangible benefits like 'social prestige and unquestioned parental authority' as well as 'stronger community support' (2005, 109). Their argument in favour of same-sex marriage relies on evidence that same-sex parents are able to reproduce the symbolic familial order of nuclear different-sex families seen as necessary for a child's 'normal' development.

Meezan and Rauch's ambivalent approach to the Oedipal process is symptomatic of the current reproblematisation of homosexuality. On the one hand, they deny that the nonheterosexuality of parents confuses the sexual development of children, a position that neither clearly denies nor affirms the standard Oedipal model, but they do not deny the importance of the basic premise of the thesis: the centrality of the affective parent-child relationship. They only disagree on the alleged outcomes of the conflict between them by suggesting that same-sex parents can, like different-sex parents, produce different-sex sexually desiring children. Moreover, by highlighting the beneficial aspects of defending 'normality' especially with regards to social reproduction (capital, mental stability, production) among families with same-sex parents, parties supporting same-sex marriage have asked the same biopolitical question as those opposing it: what forms of marriage and parenting ensure a happy, normal, socially beneficial and stable life in our society? What constitutes a happy, normal, socially beneficial and stable life is, of course, already predetermined by different-sex sexuality norms. It is no wonder then that Meezan and Rauch argue that married same-sex couples can and are likely to reproduce the social and economic relations of the legitimised heterosexual majority. From this perspective, homosexuality can reproduce the Oedipus complex, not by being the outside against which heterosexual normalcy is maintained but by reorienting the Oedipus complex to include a certain kind of homosexuality—a nonthreatening one that looks, lives, behaves, consumes and produces like heterosexuality.

In summary, the reproblematisation of homosexuality since the 1990s signals the emergence of the idea that same-sex couples can reproduce the heterosexual symbolic order and hence are not threatening to the life of the population, and therefore should not be ascribed to the realm of death. Rather, according to the new rationality that reterritorialises homosexuality under the life function, homosexuality can even serve to reproduce both the symbolic order and hence, eventually, life.

CONCLUSION

This chapter has sought to make certain conceptual observations, arguments and specifications regarding the biopolitics of life and death with regards to the apparatuses of sexuality and race. Reacting to the ontologisation of the relationship between race and death in Foucauldian scholarship, I have tried to demonstrate how strategies of life and death are not inherent to the discourses through which they are realised but rather that they are varyingly operationalised through varying discourses of difference.

The life and death functions are not historically bound and stable, nor are they fastened exclusively to the discourses of race or sexuality. They do not emanate from race and sexuality, but rather race and sexuality are strategically and tactically territorialised and reproduced by biopower in order to enact the functions of life and death. Over the past century, the strategies of biopower have shifted numerously and in complex ways, reversing, shifting and expanding its tactics to reproduce and sustain life. The life function, operating primarily through the assumption of the necessity of sexual difference for the sexual reproduction of human life, has undergone major tactical shifts since the Victorian period analysed by Foucault. Since then, sexualities once produced by the death function, such as heterosexuality and homosexuality, have been reproblematised and reterritorialised by the life function. At the crux of both tactical reversals are major changes in the sexual behaviour of Western societies; in the former it was the increase of nonreproductive different-sex sex, and in the latter, the articulation of gay liberation and emergence of new family forms. It is nothing new to regard the normalisation of homosexuality as problematic (e.g., Edelman 2004; Warner 1999), but understanding it through the life function concept adds a new dimension to understanding not only how it has emerged but also what this phenomenon is in the first place. In question is not merely the normalisation of homosexuality but its biopoliticisation as a positive (rather than a negative) apparatus of sexuality. From this perspective, the genealogical analysis of the biopolitical 'monarchy of sex' (Foucault 1981, 159) is therefore far from over.

NOTES

1. This chapter is a substantially revised version of an article previously published in *Theory & Event* under the same title (Repo 2013).
2. Foucault does not theorise his use of the term 'function', but it seems to refer to the general strategy or mode of operationalisation of power rather than, for instance, a functionalist or mathematical theorisation of biopower. In this article, I also employ the term in this sense that emphasises strategies of power.
3. It is important to note that the work of many critical race theorists challenges and/or redevelops Foucault's conception and genealogical account of race and racism (e.g., Allen 2012; Chow 2002; Goldberg 2002).
4. This is slowly changing, however, as is reflected in recent work such as in the volume by Haritaworn, Kuntsman and Posocco (2015).

BIBLIOGRAPHY

Agamben, Giorgio. 1998. *Homo Sacer: Sovereign Power and Bare Life*. Palo Alto, CA: Stanford University Press.

Allen, Theodore W. 2012. *The Invention of the White Race*, vol. 1. London: Verso.

Apostolidis, Paul. 2011. "Sex Scandals, Racial Domination and the Systematic Correlation of Power-Modalities in Foucault." *Journal of Political Power* 4 (2): 179–98.

Bernasconi, Robert. 2010. "The Policing of Race Mixing: The Place of Biopower within the History of Racisms." *Bioethical Inquiry* 7 (2): 205–16.

Butler, Judith. 1996. "Sexual Inversions." In *Feminist Interpretations of Michel Foucault*, edited by Susan J. Heckman, 59–76. Philadelphia: Pennsylvania University Press.

Butler, Judith. 2004. *Undoing Gender*. London: Routledge.

Chow, Rey. 2002. *The Protestant Ethnic and the Spirit of Capitalism*. New York: Columbia University Press.

Clarke, Victoria. 2000. "'Stereotype, Attack and Stigmatize Those Who Disagree': Employing Scientific Rhetoric in Debates about Lesbian and Gay Parenting." *Feminist & Psychology* 10 (1): 152–59.

Collins, Patricia Hill, and Sirma Bilge. 2016. *Intersectionality*. Oxford: Polity Press.

Deleuze, Gilles, and Félix Guattari. 2004. *A Thousand Plateaus: Capitalism and Schizophrenia*. London: Continuum.

Deutscher, Penelope. 2012. "Foucault's 'History of Sexuality, Volume I': Rethinking Its Reproduction." *Theory, Culture & Society* 28 (1): 119–37.

Duggan, Lisa. 2003. "The New Heteronormativity: The Sexual Politics of Neoliberalism." In *Materializing Democracy: Toward a Revitalized Cultural Politics*, edited by Dana D. Nelson and Russ Castranovo, 175–94. Durham, NC: Duke University Press.

Edelman, Lee. 2004. *No Future: Queer Theory and the Death Drive*. Durham, NC: Duke University Press.

Elbe, Stefan. 2005. "AIDS, Security, Biopolitics." *International Relations* 19 (4): 403–19.

Esposito, Roberto. 2008. *Bíos: Biopolitics and Philosophy*. Minneapolis: University of Minnesota Press.

Foucault, Michel. 1981. *The History of Sexuality 1: The Will to Knowledge*. London: Penguin.

Foucault, Michel. 2003a. *Abnormal: Lectures at the Collège de France, 1974–1975*. London: Picador.

Foucault, Michel. 2003b. *Society Must Be Defended: Lectures at the Collège de France, 1975–1976*. London: Picador.

Freud, Sigmund. 2001. *The Standard Edition of the Complete Works of Sigmund Freud*. London: Vintage.

Goldberg, David Theo. 2002. *The Racial State*. Oxford: Blackwell.

Grosz, Elizabeth. 1995. *Space, Time and Perversions: Essays on the Politics of Bodies*. London: Routledge.

Haritaworn, Jin. 2012. *The Biopolitics of Mixing: Thai Multiracialities and Haunted Ascendancies*. Farnham, UK: Ashgate.

Haritaworn, Jin, Adi Kuntsman and Silvia Posocco, eds. 2015. *Queer Necropolitics*. London: Routledge.

Hicks, Stephen. 2005. "Is Gay Parenting Bad for Kids? Responding to the 'Very Idea of Difference' in Research on Lesbian and Gay Parents." *Sexualities* 8 (2): 153–68.

Hocquenghem, Guy. 1993. *Homosexual Desire*. Durham, NC: Duke University Press.

Katz, Jonathan Ned. 2007. *The Invention of Heterosexuality*. Chicago: University of Chicago Press.

Lettow, Susanne. 2015. "Population, Race and Gender: On the Genealogy of the Modern Politics of Reproduction." *Distinktion: Scandinavian Journal of Social Theory* 16 (3): 267–82.

Loffreda, Beth. 2000. *Losing Matt Shepard: Life and Politics in the Aftermath of Anti-Gay Murder*. New York: Columbia University Press.

Mbembe, Achille. 2003. "Necropolitics." *Public Culture* 15 (1): 11–40.

McWhorter, Ladelle. 2009. *Racism and Sexual Oppression in Anglo-America: A Genealogy*. Bloomington: Indiana University Press.

Meezan, William, and Jonathan Rauch. 2005. "Gay Marriage, Same-Sex Parenting and America's Children." *Future of Children* 15 (2): 97–115.

Repo, Jemima. 2013. "The Life Function: The Biopolitics of Sexuality and Race Revisited." *Theory & Event* 16 (3). https://muse.jhu.edu/article/520029. Accessed 31 January 2007.

Repo, Jemima. 2014. "*Herculine Barbin* and the Omission of Biopolitics from Judith Butler's Gender Genealogy." *Feminist Theory* 15 (1): 73–88.

Repo, Jemima. 2016. "Thanatopolitics or Biopolitics? Diagnosing the Racial and Sexual Politics of the European Far-Right." *Contemporary Political Theory* 15 (1): 80–118.

Schultz, Susanne. 2015. "Reproducing the Nation: The New German Population Policy and the Concept of Demographization." *Distinktion: Journal of Social Theory* 16 (3): 337–61.

Smith, Anna Marie. 1994. *New Right Discourse on Race and Sexuality*. Cambridge: Cambridge University Press.

Sontag, Susan. 1990. *Illness as Metaphor and AIDS and Its Metaphors*. New York: Anchor Books.

Stacey, Judith. 1996. *In the Name of the Family: Rethinking Family Values in the Postmodern Age*. Boston: Beacon Press.

Stoler, Ann Laura. 1995. *Race and the Education of Desire*. Durham, NC: Duke University Press.

Swigonski, Mary E., Robin S. Mama and Kelly Ward, eds. 2001. *From Hate Crimes to Human Rights: A Tribute to Matthew Shepard*. New York: Routledge.

Wardle, Lynn D. 1997. "The Potential Impact of Homosexual Parenting on Children." *University of Illinois Law Review* 3: 833–920.

Warner, Michael. 1999. *The Trouble with Normal: Sex, Politics, and the Ethics of Queer Life*. Cambridge, MA: Harvard University Press.

Weheliye, Alexander G. 2014. *Habeas Viscus: Racializing Assemblages, Biopolitics, and Black Feminist Theories of the Human*. Durham, NC: Duke University Press.

Chapter 3

'Measurement of Life'

The Disciplinary Power of Racism

Hidefumi Nishiyama

INTRODUCTION: DISCIPLINE, BIOPOWER AND RACE

In *Discipline and Punish* Foucault articulates the emergence of disciplinary power in modern Europe.[1] Unlike the classical sovereign power to possess the body and to dispose it, disciplinary power, argues Foucault (1979), operates on the body through a strategy of observation and normalisation (see also Foucault 2004, 181–83). In the following year Foucault (1998; 2004) extends his inquiry of the mechanisms of modern power beyond discipline to biopower, which focuses on the biological conditions of the collective and social body as a whole. Despite differences in mode, location and historical origin, these two technologies of modern power are not, emphasises Foucault, mutually exclusive. Biopower is not a replacement of disciplinary power; instead, they cooperated and were superimposed (Foucault 1998, 146–47; 2004, 250–52). The modern government of sexuality was the quintessential example of this dual operation. As he puts it: 'Sexuality exists as the point where body and population meet' (Foucault 2004, 251–52; see also 1998, 145–46). Foucault also suggests the dual operation of modern power in his 1976 lecture *Society Must Be Defended* in the context of Nazi state racism (Foucault 2004, 259). However, this point is rather underdeveloped and he does not elaborate further how disciplinary power and biopower cooperate under the Nazi regime or modern racism in general. Reflecting this, perhaps, contemporary Foucaultian studies of race and racism tend to frame their theoretical scope of analysis predominantly in biopolitical terms. Equally, the disciplinary mechanisms of modern racism remain to be empirically explored.

This chapter examines the role of the body within the biopolitical strategy of modern racism during the years of colonialism since the late nineteenth century. I do so with an aim to reconstruct Foucault's accounts of biology, discipline, biopower and modern racism in a holistic way. My intention here is to cast light on the crucial role of disciplinary power—understood not simply as surveillance but also as an epistemological ordering of life—in the development of modern, and biopolitical, racism. Foucault (2004) theorises modern racism in relation to sociopolitical concerns with the biological conditions of 'man-as-species'. With the emergence of evolutionary thought—including, and perhaps most notably, the theory of degeneracy—the old sovereign right to take life was reinvigorated and modern racism took a pivotal role in the exercise of this power: 'The more inferior species die out, the more abnormal individuals are eliminated, the fewer degenerates there will be in the species as a whole, and the more I—as species rather than individual—can live, the stronger I will be, the more vigorous I will be. I will be able to proliferate' (Foucault 2004, 255).

More contemporary theorists, including Giorgio Agamben (1998), Robert Esposito (2008), and Achille Mbembe (2003), have extended studies of biopolitics, emphasising the power of death that is integral to the mechanisms of biopolitics and modern racism. While the nexus between sovereign power and biopower has been explicated by their works, little has been explored of the relationships among discipline, biopolitics and racism. How, if at all, does disciplinary power operate within the mechanisms of biopolitical racism? What is the role of the body—that is, 'man-as-body'—in relation to the biopolitical practice of drawing a line between populations and races? Drawing on the colonial deployments of biometrics—understood broadly and literally as 'measurement of life'—I argue that biopolitics, and its racist function in particular, does not only remain disciplinary but also is empowered by the disciplinarisation of scientific knowledges. The sovereign power to kill manifested in modern racism is, accordingly, interwoven with disciplinary power just as much as with biopolitics.

The chapter proceeds as follows: The first section revisits Foucault's account of modern racism—or the modern modality of 'race war' as he emphasises in the 1976 lectures—in the light of two correlative historical events. The first historical event is European colonialism for which Foucault is sometimes alleged to have failed to account. The second is the emergence of biology and the entrance of the notion of life into history, which Foucault articulates in *The Order of Things* (2005). The purpose of my relational analysis is not just to make an important link between the notion of life, the disciplinary power of biology—or the life sciences more generally as to include other relevant subjects such as anatomy, anthropology and pathology—and nineteenth-century race thinking. It is also to set up the context in which the surface of the body becomes a site of biopolitical configuration of populations and races, the argument which is developed in the following section.

The second section analyses the disciplinary and institutionalised observation and calculation of the body vis-à-vis race under colonialism in the late nineteenth century and the early twentieth century. The main focus is the colonial and racial deployments of biometrics or the 'measurement of life' across the world. After examining the context of Europe and during European colonialism, the section moves on to survey the development and deployment of biometrics during Japanese colonialism in East Asia. The analysis of a case of non-Western colonialism is, I suggest, of great importance for two reasons. First, it helps to highlight the role of geography in studies of race, racism and colonialism. While the biopolitical strategy of modern racism, manifested in social Darwinist theories, can be globally generalised, its codification of populations and races is always geographically specific, an aspect I have explored elsewhere (Nishiyama 2015). Second, the analysis of the 'measurement of life' also sheds light on the disciplinary power of the numerical ordering of race that exceeds the Eurocentric power-knowledge of skin colour. I argue that despite geographical variation in the mode of racism, the logic of numbers appears to have a profound impact on the global production of racial knowledge in the modern period.

COLONIALISM AND BIOLOGY: REVISITING FOUCAULT'S *SOCIETY MUST BE DEFENDED*

In the 1976 *Society Must Be Defended* lectures, as well as part five of *The Will to Knowledge*, which was originally published in the same year in French, Foucault theorises modern racism in relation to biopolitics in the nineteenth century. Foucault's theorisation of modern racism has been influential in recent decades, especially since the publication of the 1976 lectures in English, and there have been various readings of it (e.g., Bernasconi 2010; Elden 2002; Dillon 2008; Macey 2009; Mader 2011; Rasmussen 2011; Stoler 1995; Valverde 2008). This is not space here to provide a detailed exegesis of Foucault's 1976 lectures. The main purpose of this section is to reconstruct two seemingly disparate or unconnected notions in Foucault's theorisation of modern racism.

The first of these is the role of colonialism for which Foucault is sometimes alleged to have failed to account and on which postcolonial studies have expatiated. Mariana Valverde (2008, 138) argues, 'Foucault makes no attempt to theorise racism independently of the sovereignty/biopolitics framework'. 'This is important', continues Valverde, 'because as a history of racism the lectures would be very inadequate'. Postcolonial scholars have long criticised Foucault's neglect of European colonialism in his theorisation of modern racism since even before the 1997 French publication of the lectures. Foucault's

1976 lectures were first introduced to the English-speaking world in Ann Laura Stoler's *Race and the Education of Desire* (1995), which was in part a postcolonial critique of Foucault.

Departing from Foucault's account of modern racism and his history of sexuality, Stoler (1995) explores how the racial politics of colonialism was entwined with the idea of European bourgeois sexuality (see also McWhorter 2009). She (Stoler 1995) argues that biopower is intrinsically an imperial form of modern power, and suggests that the notion of 'whiteness' underlies the formation of modern European sexuality and biopolitics. Stoler, among others, shows how Foucault's theorisation of modern racism as a way of introducing a break between 'what must live and what must die' (Foucault 2004, 254)—which Foucault illustrates in the case of the Nazi regime—was deeply embedded in European colonialism and their imperial projects.

However, it should be noted that Foucault, albeit rather schematically, does acknowledge the role of colonialism in the 1976 lectures. For example, he compares the discourse of the Norman conquest during the Middle Ages with European conquests of the Americas, with reference to a parallel between William the Conqueror and Charles V (Foucault 2004, 102–103). Foucault is careful to note that it was European colonial practices abroad that were brought back to Europe since the end of the sixteenth century. He emphasises:

> It should never be forgotten that colonization, with its techniques and its political and juridical weapons, obviously transported European models to other continents, it also had a considerable boomerang effect on the mechanisms of power in the West, and on the apparatuses, institutions, and techniques of power. A whole series of colonial models were brought back to the West, and the result was that the West could practice something resembling colonization, or an internal colonization, on itself. (Foucault 2004, 103)

He also acknowledges that modern racism in the nineteenth century was 'reworked for purposes of social conservatism and, at least in a certain number of cases, colonial domination' (Foucault 2004, 65; see also 60). It was used to 'disqualify colonized subraces' (Foucault 2004, 77). At one point, Foucault (2004, 257) explicitly claims that modern racism 'first develops with colonization, or in other words, with colonizing genocide'. While none of these claims are elaborated further in the lectures, it seems clear to me that Foucault recognises the important role of colonialism for the development of biopolitical governance. Moreover, the 1976 lectures are not 'very inadequate' as a history of racism, not only because of Foucault's recognition of colonialism but also because the lectures are one of many histories of racism which happen to focus on the role of race in the mechanisms of power in a particular geographical context (namely, Europe). This is inevitable. In fact, as this chapter argues in the following section, a postcolonial study of race and racism in European colonialism is equally *a* history of racism and by no means captures complexities and multiplicities of modern racism when it was translated to another geographical context.

The second notion I would like to reconstruct is the role of discipline, the disciplinarisation of knowledges, or disciplinary power. Some commentaries suggest a period of divergence in Foucault's work in the second half of the 1970s, departing from discipline to regulation or biopolitics. David Macey (2009, 188) notes: 'Foucault [in the 1976 lectures] suddenly veers away from the 'disciplinary' notion of power elaborated in *Discipline and Punish*'. Similarly, Kim Su Rasmussen (2011, 36) argues that Foucault's genealogy of racism in *Society Must Be Defended* 'belongs to this period of transition from discipline over biopolitics to governmentality'. Accordingly, these commentaries tend to discuss Foucault's account of modern racism predominantly in biopolitical terms. In a sense, this is credible not least because Foucault himself appears to mark this transition in the lectures. Consider the following passage, the passage which Macey (2009, 188) also refers to in his reading: 'After the anatomo-politics of the human body established in the course of the eighteenth century, we have at the end of that century, the emergence of something that is no longer an anatomo-politics of the human body, but what I would call a "biopolitics" of the human race' (Foucault 2004, 243). In the last lecture, Foucault repeatedly emphasises that these technologies of power emerged in different periods.

However, as Stuart Elden (2002, 146) appositely points out, Foucault never abandons disciplinary power in *Society Must Be Defended.* There is, for Foucault, neither a separation nor a successive order, but disciplinary power and biopolitics can coexist and conjoin. Biopolitics, states Foucault (2004, 242), 'does not exclude disciplinary technology, but it does dovetail into it, integrate it, modify it to some extent, and above all, use it by sort of infiltrating it, embedding itself in existing disciplinary techniques' (see also Foucault 2004, 250). They 'were superimposed' (Foucault 2004, 249) but not replaced; they 'do not exist at the same level' but 'can be articulated with each other' (Foucault 2004, 250). 'So we have two series', says Foucault (2004, 250), 'the body-organism-discipline-institutions series, and the population-biological processes-regulatory mechanisms-State. . . . I am not trying to introduce a complete dichotomy between State and institution, because disciplines in fact always tend to escape the institutional or local framework in which they are trapped. What is more, they easily take on a Statist dimension'. Foucault takes sexuality as an example in which disciplinary power and biopower are superimposed. On the one hand, it operated on individual bodies, examining their behaviours and disciplining them; on the other hand, it operated at the population level, regulating and managing the life of the social body as a whole: sexuality is 'a matter for discipline, but also a matter for regularisation' (Foucault 2004, 251–52; see also 1998, 145–47).

Disciplinary power also operated in modern racism, although Foucault seems to give more weight to biopolitics in his 1976 lectures. It is important to note that Foucault's notion of discipline has a twofold function. On the one hand, surveillance, panopticism and the production of docile bodies play a crucial role in Foucault's theorisation of disciplinary power (1979) and have contributed to the recent development of surveillance studies (Lyon 2007). On the other hand, Foucault's notion of discipline is also about discipline in terms of the human sciences and the disciplinarisation of scientific knowledges. As Béatrice Han (2002, 116) notes, discipline here is related to Foucault's concept of power-knowledge relations (Foucault 1979, 27–28; 2001, 59); it produces 'truths' while eradicating 'false' knowledges through 'selection, normalisation, hierarchisation, and centralisation' (Foucault 2004, 181). It is, as Foucault puts it,

> [an] epistemological power, the power to extract from the individuals a knowledge and to extract a knowledge from these individuals submitted to the gaze and already controlled by different powers. . . . One can see . . . a knowledge of observation being born, a knowledge of a clinical kind, like that of psychiatry, of psychology, of psycho-sociology, of criminology. (Foucault cited in Han 2002, 118–19)

This is why, suggests Han (2002, 119), 'man' is, according to Foucault, 'an invention of the human sciences'. In other words, disciplinary power is linked to a broader epistemological structure in which individuals are subjectified and objectified, and panoptic surveillance is only one manifestation of this.

It was then biology that constituted the disciplinary power of modern racism. Modern racism does not only justify 'the death-function in the economy of biopower' (Foucault 2004, 258) but also is empowered by the disciplinarisation of scientific knowledges that allows the demarcation of populations and races and that equally contributes to its death-function. Mary Beth Mader (2011) notes an important link between Foucault's account of modern racism in *Society Must Be Defended* and his analysis of the birth of biology in *The Order of Things*. In the modern age of biology, suggests Foucault (2005), the notion of 'life' was born and came to enter the European order of knowledge. He argues that this historical event profoundly changed the way in which beings are understood and related to each other. During the classical age of natural history, natural beings were classified by their visible representation and by nominalism (Foucault 2005, 141; 144). With the historical emergence of the 'science of life' (bio-logy) around the beginning of the nineteenth century, beings became living forms and were understood in their particular relation to life (Foucault 2005, 292).

The significance of this shift is that beings were now related to each other in biological continuity; with the notion of life, evolution and genealogy were made possible. Beings were no longer considered as permanent but transformative: 'in the Classical period historical time itself . . . constituted a permanent background of continuity for natural beings that themselves were *not* essentially genealogical forms—and were not forms that could transform into other forms or types of forms, unlike, obviously, in evolutionary theory' (Mader 2011, 103). The establishment of biological continuity—particularly, evolutionary biology—was, suggests Mader (2011), the condition for the emergence of modern racism in a biopolitical form. Once beings were connected to the continuity of life in general, biological threats to species, including the concept of degeneracy, became thinkable. The function of killing under modern racism—including 'political death, expulsion, rejection, and so on' (Foucault 2004, 256)—was essentially to maintain the life of the human race as a whole. The disciplinarisation of life was thus integral to the operation of modern racism.

There is another important connection between *Society Must Be Defended* and *The Order of Things*. Stuart Elden (2002) points out a relation to the reformulated role of mathematics and calculation in the modern episteme Foucault articulates in *The Order of Things*: namely, the mathematicisation of *a posteriori* sciences, or empirical sciences, and the unification of knowledge on the basis of mathematics (Elden 2002, 137–38; see Foucault 2005, 276–78). This is evident, argues Elden (2002, 138), in thinking about race: he refers to the importance of the quantification, rather than nominal classification, of racial groups during nineteenth-century colonialism. Thus, suggests Elden (2002, 138), *Society Must Be Defended* can be understood as 'a politicizing of the argument of *The Order of Things*'. In fact, statistical measures were, for Foucault (2004, 246), a crucial mechanism of biopolitics.

The mathematicisation of life was not limited to statistical measures, however. As we see in the following section, it also prevailed within the discipline of biology during the nineteenth century. Foucault suggests this in *The Order of Things*: '[there were] a certain number of efforts that characterize modern reflection on the sciences . . . [including] the endeavour to purify, formalize, and possibly mathematicize the domain of . . . biology' (Foucault 2005, 267). He does not explore this mathematicisation of biology in greater detail in the book. Nor does he discuss it in relation to modern racism during the 1976 lectures. The next section aims at evincing this point by turning to actual disciplinary practices of measuring and objectifying the body during colonialism since the late nineteenth century. I particularly focus on practices and technology of biometrics—that is, 'measurement of life'—that were deployed at various geographical contexts, and show how the body becomes a site of biopolitical configuration of populations and races.

THE NUMERICAL ORDER OF RACE I: FROM NAZISM TO EUROPEAN COLONIALISM

Foucault briefly discusses disciplinary power in relation to modern racism with reference to Nazism:

> Of course, no State could have more disciplinary power than the Nazi regime. Nor was there any other State in which the biological was so tightly, so insistently, regulated. Disciplinary power and biopower: all this permeated, underpinned, Nazi society (control over the biological, of procreation and of heredity; control over illness and accidents too). (Foucault 2004, 259)

The Nazi state, argues Foucault, was a racist and murderous state which fully enacted the old sovereign right to kill under biological concerns. While the Nazi regime is often analysed in terms of biopower and a power of death (*inter alia* Agamben 1998), there were various ways in which disciplinary power or 'the anatomo-politics of the human body' (Foucault 2004, 243; 1998, 139) racially operated under the regime. The disciplinary power here includes surveillant and epistemological modes that were both racialised and racialising and that intertwined with the mechanisms of biopolitics. Take, for example, the inmate identification numbers that were tattooed on Jewish and Romani bodies in concentration camps. Here, the corporeal site of human flesh was used as a means of monitoring and, quite literally, of numbering, racial others, which also correlated to the management of 'other' populations. On the other hand, the body was also a site of production of racial knowledge and, in particular, of the superiority of the Aryan race over others. The manifestation of the Aryan master race was not just nominal propaganda but involved a series of scientific investigations of 'other' bodies. The body was a site of racial examination and measurement among German anthropologists, which contributed to Nazi racial policies and genocide. Even the body after death was subject to examination and display as in skeleton collection (e.g., Schafft 2004).

These disciplinary mechanisms of power in modern racism had their origins in colonialism. Indeed, they appear to be 'a boomerang effect', as Foucault puts it, of colonial models that were initially projected by the West but brought back to the West. For instance, Simone Browne (2015) argues that the surveillant use of human flesh dates back to transatlantic slavery when black bodies were branded. Also, it was British India where one of the most widely used modern methods of identification was introduced—namely, fingerprinting. The British colonial officials used fingerprinting in order to monitor natives and as civil and criminal identification (Cole 2001). Fingerprinting also has had an epistemological power since the late nineteenth century. One of the influential pioneers of fingerprint classification,

Francis Galton, who was also a cousin of Darwin and a developer of eugenics, proposed fingerprinting not just in terms of individual identification but also as a means to trace heredity. Galton hoped that all kinds of information—such as genealogy, race, criminality and intelligence, which were all important for his eugenic programme—could be identified in fingerprint patterns (Cole 2001, 99).

Although Galton himself could not realise his hope by the 1890s, similar research programmes thrived in the coming decades across Europe and the United States. As Simon A. Cole records, there were a number of scientific studies in search of 'degenerate fingerprints' during this period, including the study by Norwegian biologist Kristine Bonnevie who measured racial difference in fingerprint patterns between Europeans and Asians and concluded that the former were more evolved than the latter (Cole 2001, 111). Joseph Pugliese also analyses various scientific measurements of the body since the late nineteenth century and demonstrates the biometric construction of race, in particular, that of 'the template white body' (Pugliese 2010, 30) in which the normalising power of whiteness is manifested while excluding and pathologising nonwhite bodies. Pugliese also suggests the twofold operation of disciplinary power and biopower in French police anthropometry: 'police anthropometry . . . focused on the exercise of disciplinary power on the body-organism of the individual in the institutional context of the police station *and* on the larger, biosocial problem of the state in identifying and regulating criminal populations of "recidivists"' (Pugliese 2010, 53).

THE NUMERICAL ORDER OF RACE II: JAPANESE COLONIALISM IN EAST ASIA

The numerical racial ordering of the body was, however, more extensive and permeating and therefore must be explored beyond Eurocentrism and the realm of whiteness. In writing a history of racism, there seems to be a tendency to focus exclusively on the European, and more broadly Western, (colonial) context—a tendency which allows reducing the problematic of racism to whiteness. Reflecting Cornel West's genealogy of modern racism which focuses on European powers and skin colour, Ian Hacking suggests 'rethinking the connection between race and geography' and its link to *each* empire since the '[c]lassification of peoples by a category of race is an integral part of the control necessary to organized and maintain an empire' (Hacking 2005, 110; 116).

In fact, contrary to W. E. B. Du Bois's famous dictum, 'the problem of the twentieth century is the problem of the color-line' (Du Bois 2007, 15), skin colour is only one form of manifestation (and materialisation) of race.

As Michael Weiner puts it, 'emphasis on the "Colour Line", as conceptualized by Du Bois, runs the risk of reifying skin colour—of ignoring the fact that the visibility of somatic difference is itself a social construct' (2009, xiv). This is evident in the cases of racism against Koreans, the Chinese, and other Asians in Japan, racism against the Irish in the United Kingdom, and, not to mention, racism against Jews in Europe. Racism in Japan in particular can be assimilated neither by the colour line nor by whiteness because it is, unlike the other cases, the Japanese themselves who engage with practices of racialisation against non-Japanese Asians.

The scientific logic of numbers and the mathematicisation of race prevailed over the dominant racist discourse of pigmentation. During Japanese colonialism in East Asia, anthropological and biological research on 'other' bodies began to flourish after the colonisation of Taiwan in 1895 and was widely conducted throughout the early twentieth century. In the early years of Japanese occupation in Taiwan, Baron Gōtō Shimpei, chief of the Civil Administration Bureau of Taiwan, proposed what he called 'colonial governance by biological principles' (Takekoshi 1905, 36). In his book *A Theory of Japanese Colonialism* (*Nihon shokuminchi ron*), Gōtō (1915, 22–25) explained that colonial governance by biological principles requires extensive research on the living conditions of the natives, natural phenomena and resources, and so forth, and implements colonial policies according to these conditions. The Japanese colonial strategy also entailed the foundation of an anthropological research institute which aimed at examining and knowing colonial bodies. During the last months of the First Sino-Japanese War, Tsuboi Shōgōrō (1895), the leading anthropologist of that time, urged to set up the study of race, and the Taiwan Anthropological Society was established in Taipei. Japanese anthropologists in the colony began to investigate 'other' bodies, particularly aboriginal tribes or 'raw barbarians' (*seiban*), the term that was generically used during the earlier years of the occupation.

The knowledge of colonial bodies was an integral part of colonial domination. As Japanese anthropologist Mori Ushinosuke (cited in Tierney 2012, 129) put it: 'If we are to subjugate the aborigines, we must of course know them'. Since the early years of occupation, anthropologists such as Inō Kanori and Torii Ryūzō observed and measured a biological constitution of each tribe by using various anthropometric measurements. Their studies were published in the *Journal of the Anthropological Society of Tokyo* (*Tokyo jinrui gakkai zasshi*), the predecessor of the *Journal of Anthropological Society of Nippon* (*Jinruigaku zasshi*) (e.g., Inō 1898; Mori 1914; Torii 1904).

A similar project on colonial bodies also took place in Korea after the 1910 annexation. Kubo Takeshi (1915, 122), a professor of anatomy at a newly reorganised, colonised (in 1915), medical institution Keijō Medical College (*Chōsen sōtokufu keijyo igaku senmon gakkō*), proposed what he called 'anatomical anthropology of the Koreans'. Kubo (1915), who had also been researching in Korea for four years prior to the annexation, analysed an anatomo-biological constitution of the Korean body and compared it with that of the Japanese, using extensive measurements ranging from muscle and internal organs to the size of penis. For Kubo, like Mori above, the knowledge of colonial bodies was an integral part of colonial governance. He proclaimed that it is vital for ethnic governance to study a biological constitution of each ethnic group and identify superiority and inferiority among different groups. This, continued Kubo, allows the Japanese colonial power to educate the Koreans and improve their inferior parts (Kubo 1919, 70).

The colonial will to knowledge under the Japanese empire was closely linked to the notion of criminality and degeneracy within an evolutionary framework. One of the events that clearly manifested this was the so-called Kubo incident. In May 1921, when a skull went missing from a laboratory at Keijō Medical College, 'Kubo accused the Korean students of the crime based on his experience in the physical anthropology of the Korean people. Anatomically speaking, Kubo elaborated, Koreans were barbarians whose racial traits determined their historical developments' (Kim 2013, 412). Similarly, Miyake Hideo, an anatomist at Nanman Igakudō in Manchuria, conducted an anatomical study of Korean bodies with particular reference to their palm prints and fingerprints. Comparing them with Polish samples, Miyake (1922, 761) concluded that Korean palm prints, which appear to be the same as Chinese samples, signify their degeneracy in comparison with Poles. Moreover, the discourse of degeneracy and barbarism was particularly prominent in studies of Taiwanese aborigines. Kanaseki Takeo (1929, 520), a professor at Taihoku Imperial University in the 1930s, collected biometric data of 'raw barbarians' for the purpose of 'statistical observation'. Classified into racial groups including Atayal (a Taiwanese aboriginal tribe), Japanese, Korean, Chinese and Europeans, Kanaseki (1929, 544) asserted that the fingerprints of Atayal men reveal that they are unduly primitive compared to other neighbouring races.

Among various measurements of the body, fingerprinting became a widespread technology of racial knowledge during Japanese colonialism especially after the introduction of the 'fingerprint index' by Furuhata Tanemoto. Born in 1891, Furuhata was a professor of forensic medicine at Kanazawa Medical University and later at Tokyo Imperial University and one of the leading figures in the development of forensic science in the country. In his 1926 lectures at the Kanazawa Association of Criminology and the Hokuriku Association of Medical Sciences, Furuhata proposed the biological application of fingerprinting which he called 'fingerprint index' (*shimon keisū*). Furuhata's fingerprint index was a scientific method of reading heredity and race through anatomical classification and statistical calculation of fingerprint patterns. It was based on earlier European studies of fingerprints, including Galton's classification of arches, loops and whorls.

Furuhata (1926) introduced a mathematical formula which was calculated by dividing the proportion of whorls by the proportion of loops (both radial and ulnar) and then multiplied by 100: *fingerprint index = Whorls/Loops × 100.* He calculated fingerprints of Europeans and Asians and propounded their racial hierarchical order: he asserted the lower fingerprint index among Europeans—generally below 70—in contrast to the higher fingerprint index among Asians, including Manchurians, Chinese, and Taiwanese 'raw barbarians', some of which were beyond 100 (Furuhata 1926, 5). Furuhata (1926, 6) also related his fingerprint index to craniometry, claiming that it tends to be lower among the brachycephalic people, while it tends to be higher among the dolichocephalic people. Based on his study, Furuhata suggested a categorisation into four types: a score of over 90 was classified as the Manchurian type; between 90 and 70 as the Japanese type; between 70 and 60 as the Italian type; between 60 and 50 as the Indian type; and below 50 as the Western type (Furuhata 1926, 7).

This simplified and generalisable mathematical formula of the fingerprint index was widely used in the following years by Furuhata's students and colleagues, as well as other advocates, including biologists, anatomists and anthropologists. It was followed by large-scale studies, using fingerprint samples varying from Europeans (including British, Italian, Norwegian, Polish, and German) and Asians (Chinese, Korean, Indian, etc.) to indigenous and colonised populations in East Asia (Ainu, Ryukyuan, and Taiwanese 'raw barbarians'). The scope of fingerprint index calculation was predominantly concerned with the knowledge of race in East Asia in terms of both the 'Japanese race' and its racial others. One of the earlier studies was conducted within the mainland of Japan as an attempt to articulate a regional difference and distribution of fingerprint patterns and to determine the biogeographical origin of the Yamato people, which was believed to be the dominant native population in the history of Japan.

In the second issue of the 'Research on Japanese Fingerprint' (*Nihonjin shimon no kenkyū*) series, Hirai Sumimaro (1928) conducted a study to understand 'racial peculiarities' and to investigate the origin of the Yamato people. Hirai analysed Japanese fingerprint samples across all forty-seven prefectures, and suggested the tendency that the fingerprint index is higher in prefectures on the coast of the Sea of Japan while it is lower on the side of the Pacific. The result was also compared with other biological measurements—what Hirai referred to as 'race index' (*jinshu keisū*) that was based on blood type and the measurement of height and head. The study concluded that there is consistency among these three types of biological data which can help to identify the biogeographical origin of the Yamato people in relation to the influence of other Asian races, suggesting a possible racial influence of non-Yamato populations from the continent in history (Hirai 1928; see also Furuhata 1930, 285).

Tracing the hereditary origin of the Yamato people, the early biometric reading of the body was incorporated into the search of the 'pure' Japanese race. The fingerprint index became a technology of purifying the Japanese race in the sense that the search for the origin through fingerprint patterns was constitutive of what was later understood as pure Japanese. It was not about the purification of race in terms of the elimination of the abnormal and degenerate as manifested in modern state racism in Europe (Foucault 2004), but nevertheless a technology of racial normalisation that operated on the surface of the body.

The internal construction of the Japanese race was coupled with the calculation of non-Japanese races. In subsequent years, studies of fingerprint patterns became increasingly about the production of racial knowledge in East Asia whereby other Asian races were not simply biologically differentiated but also hierarchised. Fingerprint data collected from mainland Japan, its colonies and beyond were all put into the fingerprint index formula and racial and national groups across the world were systematically ordered. Table 3.1 is an extract of the 'fingerprint classification table by race', published in 1935 in the series of the Research on Japanese Fingerprints (Hibino 1935). It ranges from the high fingerprint index of some of Taiwanese 'raw barbarians'—170.18 of Atayal and 127.61 of Ami—to the lowest fingerprint index of Jews, 30.56. The number of the fingerprint index was understood as the level of civilisation and higher indexes were interpreted as the indicator of barbarism and degeneracy. Another study of the 'Research on Japanese

Table 3.1 Fingerprint classification table by race

Non-Japanese	*Fingerprint index*	*Japanese*	*Fingerprint index*
Atayal (Taiwanese aborigine)	170.18	Shiga	107.52
Ami (Taiwanese aborigine)	127.61	Ishikawa	102.71
Yami (Taiwanese aborigine)	108.50	Gifu	99.83
Cantonese	108.25	Toyama	98.58
Korean	94.37	Ishikawa	93.30
Ryukyuan	89.01	Kumamoto	92.28
Tsuo (Taiwanese aborigine)	86.96	Ehime	89.49
Italian	62.38	Shizuoka	85.33
American	51.19	Nigata	81.34
Paiwan (Taiwanese aborigine)	48.29	Fukui	80.15
German	46.15	Kyoto	78.76
Ainu	42.37	Toyama	74.26
British	36.51	Tokyo	72.94
Jews	30.56		

Fingerprints' conducted a comparative analysis of five Taiwanese aboriginal tribes, namely, Paiwan, Tsuo, Yami, Ami, and Atayal, and calculated their fingerprint indexes as 51.31, 86.96, 107.93, 127.78, and 161.64, respectively (Itō 1935, 1441). The study concluded the biologically inherited barbarity of Atayal, while suggesting the relatively civilised bodies of Paiwan and Tsuo—the tribes that were seen as more obedient under the Japanese rule.

The fingerprint index studies and colonial anthropological research more broadly during the period of Japanese imperial rule exhibits a disciplinary role in the biopolitical configuration of populations and races, which was geographically coded. Like earlier studies in Europe since the late nineteenth century, examination of the body during Japanese colonialism became a technology of racialisation, demarcating between 'normal' or 'civilised' fingerprints and 'abnormal' or 'barbarous' fingerprints. Yet, importantly, this was not so much in accordance with the European taxonomy of race or racial pigmentation but in relation to Japan's own colonial governance in East Asia. It was to establish its own distinct raciality and its relative superiority to other Asians or other 'yellows'. While the Eurocentric and white supremacist mode of scientific racialisation never totally ceased during Japanese colonialism, the concept of race and practices of racialisation were irreducible to it; instead, they were locally appropriated. The nineteenth-century European idea of degenerate fingerprints was multiplied; instead of challenging the 'degenerate fingerprints' of Asians proposed by European scientists, other degenerate fingerprints were constructed across East Asia.

That is to say, the power-knowledge of biometrics emerges as spatially specific, intertwining the knowledge of the body with imperial power in each geographical and historical context. Equally, the disciplinary power of modern racism operated in a geographically and historically specific manner. While the logic of biopolitical racism may be—and, in fact, has been—globally generalised, it is such localised disciplinary power that empowered each colonial and racist form of biopolitics.

CONCLUSION

The mathematicisation of biology appears to have contributed to the production of racial knowledge since the late nineteenth century, not only in the West and its colonial context but also in a non-Western mode of colonialism. This is not to say that the mathematicisation of biology conditioned the invention of the modern concept of race, but it nevertheless fortified modern racism in a scientific form across imperialisms. Drawing on the cases of colonial deployment of the measurement of life, the chapter has explored some of the ways in which the body was utilised as a domain of exercising disciplinary, (racially) normalising and (colonial) power-knowledge practices. Producing racial knowledge, the surface of the body became a site of biopolitical configuration whereby bodies were ordered by a series of evolutionary notions of degeneracy, barbarism and civilisation. At the same time, I have attempted to reconstruct and integrate Foucault's accounts of biology, discipline, biopower and modern racism in the context of colonialism. Although the link between his concepts is not always explicit and some of his points are left underexamined, a holistic reading of Foucault is not only plausible but also helps to understand the mechanisms of biopolitical racism beyond the problematic of whiteness.

An analysis of the disciplinary power of modern racism also cautions against a universalist approach to studies of biopolitics. The logic of biopolitical thinking may be translated across different contexts. Biopolitics and its death function are, however, neither ahistorical nor aspatial. As the chapter has shown, both the modern disciplinarisation of scientific knowledges and a localised mode of the production of knowledge are integral to the operations of biopolitical racism. In fact, an analysis of disciplinary power of modern racism, examining the production of local knowledges, allows us to understand historical and geographical transformations of the way in which biopolitical racism is empowered. Disciplinary power does not only continue to perform in biopolitics but also plays a part in the reinvigoration of the right to kill under the biopolitical theme through the production of the scientific 'truth' about races. If there is a continuity between sovereign power and biopower in terms of drawing a line between 'what must live and what must die', as Agamben (1998) suggests, so is there the continuity of disciplinary power. Yet, disciplinary power continues in biopolitics as much as reorganises the biopolitical order of populations and races according to each imperial context.

NOTES

The chapter develops arguments originally made in my article (Nishiyama 2015). I would like to thank the Academy of Finland for funding (RELATE CoE, grant 307348) that allowed me to continue this research.

1. There is inconsistency in Foucault's use of the terms *disciplinary power*, *biopower* and *biopolitics*. In *The Will to Knowledge* Foucault seems to suggest that biopower consists of disciplinary power and biopolitics: they are 'two poles'—'*the disciplines: an anatomo-politics of the human body*' and '*regulatory controls: a biopolitics of the population*'; they are 'two directions' in 'an era of biopower' (Foucault 1998, 139–40). In the 1976 lectures, in contrast, Foucault appears to differentiate between disciplinary power and biopower when he discusses the historical emergence

of each technology of power. For example, he articulates the emergence of 'the technology of biopower, of this technology of power over "the" population as such' that is no longer 'disciplined' but 'regularized' (Foucault 2004, 247). In this chapter, I use the terms in the latter sense: disciplinary power, and Foucault's account of discipline, is not part of biopower but they are two different, yet interrelated, technologies of power; and biopolitics as a form of political formation and organisation of the state underpinned by biopower, that is, the technology of power directed at and regulates the biological conditions of the social body.

BIBLIOGRAPHY

Agamben, Giorgio. 1998. *Homo Sacer: Sovereign Power and Bare Life*. Palo Alto, CA: Stanford University Press.

Bernasconi, Robert. 2010. "The Policing of Race Mixing: The Place of Biopower within the History of Racisms." *Bioethical Inquiry* 7 (2): 205–16.

Browne, Simone. 2015. *Dark Matters: On the Surveillance of Blackness*. Durham, NC: Duke University Press.

Cole, Simon A. 2001. *Suspect Identities: A History of Fingerprinting and Criminal Identification*. Cambridge, MA: Harvard University Press.

Dillon, Michael. 2008. "Security, Race and War." In *Foucault on Politics, Security and War*, edited by Michael Dillon and Andrew W. Neal, 166–96. London: Palgrave Macmillan.

Du Bois, W. E. B. 2007. *The Souls of Black Folk*. Oxford: Oxford University Press.

Elden, Stuart. 2002. "The War of Races and the Constitution of the State: Foucault's (Il Faut Défendre La Société) and the Politics of Calculation." *Boundary 2* 29 (1): 125–51.

Esposito, Robert. 2008. *Bíos: Biopolitics and Philosophy*. Minneapolis: University of Minnesota Press.

Foucault, Michel. 1979. *Discipline and Punish: The Birth of the Prison*. Middlesex: Penguin Books.

Foucault, Michel. 1998. *The Will to Knowledge: The History of Sexuality*, vol. 1. London: Penguin.

Foucault, Michel. 2001. *Madness and Civilization: The Birth of the Clinic*. London: Routledge.

Foucault, Michel. 2004. *Society Must Be Defended: Lectures at the Collège de France, 1975–76*. London: Penguin.

Foucault, Michel. 2005. *The Order of Things: An Archaeology of the Human Sciences*. London: Routledge.

Furuhata, Tanemoto. 1926. "Shimon jinshū tokuisei narabini 'shimon keisū' ni tsuite." *Nippon no ikai* 16 (93): 5–7.

Furuhata, Tanemoto. 1930. *Ketsuekigata to oyako kantei, Shimon-gaku*. Tokyo: Bukyōsha.

Gōtō, Shimpei. 1915. *Nihon shokuminchi ron*. Tokyo: Kōmindōmei shuppanbu.

Hacking, Ian. 2005. "Why Race Still Matters." *Daedalus* 134: 102–116.

Han, Béatrice. 2002. *Foucault's Critical Project: Between the Transcendental and the Historical*. Palo Alto, CA: Stanford University Press.
Hibino, Masaru. 1935. "Nihonjin shimon no kenkyū 17: Ryūkyū jin shimon no kenkyū." *Jyūzenkai Zasshi* 40 (6): 2189–2202.
Hirai, Sumimaro. 1928. "Nihonjin shimon no kenkyū 2: 1, shimon keisū no jinruigakuteki ōyo ni tsuite." *Jyūzenkai Zasshi* 33 (9): 1255–72.
Inō, Kanori. 1989. "Taiwan ni okeru kaku banzoku no bunpu." *Tokyo jinrui gakkai zasshi* 146: 301–307.
Itō, Shizuo. 1935. "Nihonjin shimon no kenkyū 16: Tsuo zoku no shimon ni tsuite." *Jyūzenkai zasshi* 40 (4): 1438–51.
Kanaseki, Takeo. 1929. "Seiban jin teashi hihu no rimon ni tsuite 1." *Jinruigaku zasshi* 44 (11): 519–46.
Kim, Hoi-Eum. 2013. "Anatomically Speaking: The Kubo Incident and the Paradox of Race in Colonial Korea." In *Race and Racism in Modern East Asia*, edited by Rotem Kowner and Walter Demel, 411–30. Leiden: Brill.
Kubo, Takeshi. 1915. "Chōsenjin no taishitsu teki jinruigaku hoi." *Jinruigaku zasshi* 30 (4): 122–37.
Kubo, Takeshi. 1919. "Chōsenjin no jinshu kaibō gakuteki kenkyū." *Jyūzenkai zasshi* 24 (2): 69–90.
Lyon, David. 2007. *Surveillance Studies: An Overview*. Cambridge: Polity.
Macey, David. 2009. "Rethinking Biopolitics, Race and Power in the Wake of Foucault." *Theory, Culture & Society* 26 (6): 186–205.
Mader, Mary Beth. 2011. "Modern Living and Vital Race: Foucault and the Science of Life." *Foucault Studies* 12: 186–205.
Mbembe, Achille. 2003. "Necropolitics." *Public Culture* 15 (1): 11–40.
McWhorter, Ladelle. 2009. *Racism and Sexual Oppression in Anglo-America: A Genealogy*. Bloomington: Indiana University Press.
Miyake, Hideo. 1922. "Chōsenjin no te no rimon." *Nanman igaku dō ronshō* 11: 754–69.
Mori, Ushinosuke. 1914. "Taiwan no seiban ni tsuite." *Jinruigaku Zasshi* 29: 54–60.
Nishiyama, Hidefumi. 2015. "Towards a Global Genealogy of Biopolitics: Race, Colonialism, and Biopolitics beyond Europe." *Environment and Planning D: Society and Space* 33 (2): 331–46.
Pugliese, Joseph. 2010. *Biometrics: Bodies, Technologies, Biopolitics*. London: Routledge.
Rasmussen, Kim Su. 2011. "Foucault's Genealogy of Racism." *Theory, Culture & Society* 28 (5): 34–51.
Schafft, Gretchen E. 2004. *From Racism to Genocide: Anthropology in the Third Reich*. Champaign: University of Illinois Press.
Stoler, Ann Laura. 1995. *Race and the Education of Desire: Foucault's History of Sexuality and the Colonial Order of Things*. Durham, NC: Duke University Press.
Takekoshi, Yosaburō. 1905. *Taiwan tōji shi*. Tokyo: Hakubunkan.
Tierney, Robert. 2012. "Violence, Borders, Identity: An Ethnographic Narrative Set in Colonial Taiwan." In *Reading Colonial Japan: Text, Context, and Critique*, edited by Michele M. Mason and Helen J. S. Lee, 124–40. Palo Alto, CA: Stanford University Press.

Torii, Ryûzô. 1904. "Chosenjin no taishitsu." *Jinruigaku Zasshi* 20 (25): 142–48.

Tsuboi, Shōgorō. 1895. "Jinshu mondai kenkyū no jyunbi." *Tokyo jinruigakkai zasshi* 108: 214–22.

Valverde, Mariana. 2008. "Law versus History: Foucault's Genealogy of Modern Sovereignty." In *Foucault and Politics, Security and War*, edited by Michael Dillon and Andrew Neal, 135–50. London: Palgrave Macmillan.

Weiner, Michael. 2009. "Editor's Introduction." In *Japan's Minorities: The Illusion of Homogeneity*, 2nd ed., edited by Michael Weiner, xiv–xxii. Oxon, UK: Routledge.

Part II

MAPPING INTERSECTIONAL GEOGRAPHIES OF THE BODY: RACE, GENDER, SEXUALITY, ECONOMY

Chapter 4

Homo Sacer Is Syrian

Movement-Images from the European 'Refugee Crisis'

Hannah Richter

THE AGAMBENIAN COLOURING OF THE EU 'REFUGEE CRISIS'

Since early 2015, debate and discussion around the arrival of large numbers of refugees on European territory has become a stable part of the political discourse throughout the European Union (EU). In April 2015, it prompted right-wing UK columnist Katie Hopkins to declare that she, indeed, does not care. 'Show me pictures of coffins, show me bodies floating in water, play violins and show me skinny people looking sad. I still don't care' (Hopkins quoted in Martin 2015). Hopkins was ultimately forced to apologise for her suggestion to use gunships to deter migrants. But her statements reveal the vast discrepancy between the dominant political framing of refugee displacement as 'the world's largest humanitarian crisis since World War II' (EC 2016a), the parallel governmental securitisation of 'fortress Europe' not far removed from Hopkins's suggestions, and the contempt of the political right for what they identify as the naïve blindness of do-gooders towards the exploitation of Western economies and social security systems by rationally calculating bogus asylum seekers (Kmak 2015, 404–407).

However, Hopkin's mocking words also give insight into what the polarised political framing of the EU 'refugee crisis'[1] is centred on: media images which have come to symbolise the migration movement in its extraordinary humanitarian dimension (the body of two-year-old Alan Kurdi on the Turkish shore, for example) and the political necessity to address the former (the selfie which the Syrian Anas Modamani took with German chancellor Angela Merkel) (Mortensen 2016, 415–19). But at the same time, popularised images also illustrate the spectacular securitisation of European as well as national borders (such as the destruction of the Calais 'jungle camp') (BBC 2016) in the face of exceptional migrant numbers: in both 2015 and 2016, at the peak of the crisis, an unprecedented 1.3 million people crossed European borders and applied for asylum in the EU, highly varied in terms of their origin, but with Syrian, Afghan and Iraqi citizens recorded as the largest nationality groups (Eurostat 2017).

Drawing on the themes of exceptionalism as well as inclusion/exclusion implicit here, this chapter unpacks the institutional-political reaction to the European 'refugee crisis' through the theoretical lens provided by Giorgio Agamben's *Homo Sacer* series. Agamben, who 'in the past decade . . . has challenged Hannah Arendt's place as the "charismatic legitimator" within critical Refugee Studies' (Owens 2009, 567), theorises the refugee as the bodily site of the indistinction which holds the 'ontologico-political machine' (Agamben 2016, 239) of biopolitics in motion. The inclusion of extraordinary political measures under emergency circumstances allows sovereign power to reproduce itself from the void excluded from its ordinary governance. Agamben describes this exceptional mechanism of 'inclusive exclusion' (1998, 22) with a reference to Gilles Deleuze and Félix Guattari's *A Thousand Plateaus*: 'Sovereignty only reigns over what it is capable of internalizing' (1987, 360; as quoted in Agamben 1998, 18). But the sovereign machine is biopolitical. Its operationality is centred on the nodal point of a politically constituted pre- or extra-political bare life which 'dwells in the no-man's-land between the home and the city' (ibid., 90).

This included negative of bare life is embodied by the refugee as *Homo sacer* par excellence. The refugee can be killed but cannot be sacrificed—she is excluded, but this exclusion is a necessary part of the reproductive mechanism of sovereign power. 'When the rights of man are no longer rights of the citizen, he is truly *sacred,* in the sense that this term had in archaic Roman law: destined to die' (Agamben 2004, 117). In the following, I show that Agamben's theory provides a useful lens to unpack European governmental reactions to the increased arrival of asylum seekers publicly framed as 'refugee crisis'. But from this Agambenian perspective, state responses which go beyond the undifferentiated exceptional containment of refugee movement, such as the initially comparatively welcoming approaches in Germany and Sweden, remain puzzling.

The one-dimensionality of Agamben's political figure of the refugee is linked back to the static spatiality which problematically underlies both his theorisation of sovereign power and of resistance. I therefore suggest supplementing Agamben's conceptualisation of the refugee as *Homo sacer* with Deleuze and Guattari's more open territoriality of sovereignty. Beyond the exceptional stasis which grounds the sovereign realm in Agamben, I draw on

Deleuze's cinematic movement-image as an analytical tool to understand the politically reproductive employment of migration movement for a governance which is dynamic, aligning and framing. But Deleuze's movement-images always entail the potential to actualise a multiplicity of spatio-temporal lineages if opened up beyond the closed system of a particular frame. As multifaceted time-images, they can form the basis to link refugee movement and resistance in a way which does not idealise the politically inoperative, liminal status of the refugee, as it is the case in Agamben.

EXCEPTION AND EXCLUSION: THE TWO FACES OF AGAMBEN'S ONTOLOGICO-POLITICAL MACHINE

In the political reaction to the EU 'refugee crisis', the two intrinsically related dimensions of political exceptionalism and spatial-legal exclusion from the realm of sovereign governance are clearly identifiable. As a political event, the refugee crisis constitutes a state of exception in its original sense. It is identified as a threat to the security of the state and thus allows for the momentous, constitutive extension of sovereign power beyond the realm of ordinary legality. Developed by Agamben in the early volumes of his *Homo Sacer* series, the state of exception resolves the problem of self-presupposition faced by a political sovereign whose authority lacks ontological grounding and is only constituted through its exercise. Unbound from the necessity to follow existing legal rules because they provide no guiding precedence for the extraordinary situation, the exception allows sovereign governance to momentarily transcend itself and thereby to 'define the normal case as the realm of its own validity' (Agamben 1998, 17). In the state of exception, sovereign politics can become its own constitutive outside because it 'divides itself into constituting power and constituted power and maintains itself in relation to both, positioning itself at their point of indistinction' (Agamben 1998, 41).

The characterisation of recent EU migration as an exceptional case so particular that regular political measures cannot apply here becomes evident in the way it is referred to by European political leaders. In August 2015, German chancellor Angela Merkel defined 'the issue of asylum' as 'the next major European project' which would 'preoccupy Europe much, much more than the issue of Greece and the stability of the Euro' (Merkel quoted in Berry, Garcia-Blanco, and Moore 2015). With even more urgency, the president of the European Council, Donald Tusk, identified the EU 'refugee crisis' as a serious challenge to the sociopolitical status of the EU which must be tackled immediately, because 'time is running out' (Holehouse and Fraser 2015). Official publications by EU bodies and international organisations emphasise the unprecedented character of the recent migration into Europe, in both quantity and character, as a political task for Europe which requires urgent responses to re-establish political control (EC 2016b; Spindler 2015; Spijkerboer 2016, 548–50).

Without downplaying the seriousness of the recent refugee movements both for the people displaced and as a political challenge, several researchers question the necessity of its exceptional political framing (Goddeeris 2016; Spijkerboer 2016, 534–36; Alcalde 2016). They point to the mass flight of Hungarians after the failed uprising against Soviet rule in 1956 or the displacement of Bosnians during the Balkan wars to illustrate that, while the number of refugees recorded in 2015 and 2016 was twice as high compared to earlier peaks (Eurostat 2017), the situation in the EU—at this time only fifteen member-states—was very similar, making the current migration 'hardly unprecedented' (Kennedy 2016). This suggests that the presentation of recent migration movements as a crisis, threat and unprecedented challenge is not only politically contingent but also indefinitely politically productive. As Claudia Aradau and Rens van Munster argue in *Politics of Catastrophe*, the idea of a 'crisis' evokes an exceptionality which is ahistorical; mobilising 'forms of knowledge that oscillate between the probable and the possible' (2011, 23), the necessity for and legitimacy of exceptional political means is extended indefinitely in the face of an epistemologically and politically insecure future.

But the politically constitutive use of exceptional means also underlies the extended border securitisation with which European countries reacted to the arrival of increasing numbers of refugees, and is illustrated in the multiple images it produces. In Hungary, a state of emergency was declared on 15 September 2015. When executive forces were no longer able to secure the border to Serbia, the Hungarian government declared all instances of border crossing as illegal immigration and opened them up to prosecution (BBC 2016). In summer 2015, several EU countries such as Austria, Denmark or Germany temporarily suspended transnational transportation links as well as the Schengen agreement to reinstate border controls and prevent refugees from entering their national territory (BBC 2016). But the transcendent, exceptional violence through which the political sovereign reproduces its governmental realm is perhaps most vividly illustrated by the Sisyphean task of the French border guards who prevent the inhabitants of the Calais 'jungle camp' from entering the United Kingdom night after night, up to the destruction of the camp in October 2016 (BBC 2016; Rygiel 2011, 7–13).

Different from the usual clandestine outsourcing of border violence to the countries at the margins of Europe, the explicit and mediatised display of force within here constitutes a 'border spectacle' (De Genova 2013, 1181) at the heart of Europe (Vaughan-Williams 2015; Diken 2004). With Agamben, this border spectacle—played out on the violently excluded bodies of the migrants—can be understood as the dramatic enactment of a transcendent political force which lacks immanent ontological grounding.

While Agamben sets out to explore how sovereign governance makes use of exceptional transcendence to renew its claim to political sovereignty at the beginning of his *Homo Sacer* project, his analytical focus shifts in the third book of the series, *The Kingdom and the Glory*. Pointing out the analogous functionality of theological authority and political power, Agamben now reveals how a ceremonial staging of force is necessary to bring this claim into existence in the first place. Both theological and political-legal authority rely on the immanent primacy of an Aristotelian unmoved mover without ontological essence. For this reason, their immanence must retroactively be brought into existence through an economy of ritualised, transcendent enactment. 'There is no substance of power, but only an "economy", only a "government"' (Agamben 2011, 139). The border spectacle is thus a substitute for the ceremonial glory of the king deemed suitable for display in the context of contemporary Western democracies. Both conceal the empty throne, the groundlessness of governmental authority, through the staging of forceful political action (ibid., 188–96).

But the allusion to Guy Debord is not accidental here. In *The Use of Bodies,* Agamben argues that just like Debord's society of the spectacle only functions against the background of a private life kept secret, the apparatus of immanent sovereign power and transcendent exceptional force reproduces itself dialectically between *zoē* and *bios* (Agamben 2016, xv–xix). The constitutive transition between both poles of the biopolitical economy, however, requires a liminal zone of indistinction both exempt from the political-legal framework of sovereign rule and included in its ordinary functioning. For Agamben, the camp is this nodal point of biopolitics, 'the pure, absolute, and impassable biopolitical space' (Agamben 1998, 123). The camp is the spatial embodiment of 'the becoming-rule of exception' (Diken 2004, 96) exemplified by the concentration camps of the Nazi government. *Homo sacer,* the liminal figure politically situated in the zone of indistinction between *zoē* and *bios*, is its natural inhabitant.

> The bare life into which the camp's inhabitants were transformed is not, however, an extrapolitical, natural fact that law must limit itself to confirming or recognizing. It is, rather, a threshold in which law constantly passes over into fact and fact into law, and in which the two planes become indistinguishable. It is impossible to grasp the specificity of the National Socialist concept of race—and, with this, the peculiar vagueness and inconsistency that characterize it—if one forgets that the biopolitical body that constitutes the new fundamental political subject is neither a *quaestio foeti* (for example, the identification of a certain biological body) nor a *quaestio iuris* (the identification of a certain juridical rule to be applied), but rather the sire of a sovereign political decision that operates in the absolute indistinction of fact and law. (Agamben 1998, 171)

In the camp, the law of the emergent sovereign Leviathan gives itself not one but several bodies to 'pass over into fact', physically endow itself with the substance it ontologically lacks. To do so, it legally neutralises the agency and identity of the *homines sacri* whose bodies it borrows, stalling the political force of their movement through spatial confinement. In the governance of the EU 'refugee crisis', this political use of the camp as a space of both physical confinement and desubjectification, normalised as the visible constituent of the self-reproductive machine of sovereign power, is clearly evident. Following the declaration of the state of exception in Hungary, refugees were forcefully moved to registration camps. Broadcasted by the media were disturbing images of people clinging to rails, chanting: 'No camps, no camps' (BBC 2016). Especially Greece and Italy have been subject to sustained criticism for the exceptional incarceration of arriving refugees without legal charge, their geographical but also political exclusion from the European political community and the rights guaranteed by its citizenship which Dimitra Kotouza further illustrates in her chapter in this collection (Vaughan-Williams 2015; Ticktin 2005; Rygiel 2011).

But more generally, the focused political attempt to house arriving refugees away from urban areas and the infrastructure they offer in many EU states, particularly the restriction of their free movement in Germany, can be understood as the political use of the camp in Agamben's sense, the governmental inclusion of refugees as spatially excluded *homines sacri* (Diken 2004, 92–95). In his identically titled commentary on Hannah Arendt's *We Refugees*, Agamben, however, introduces a second dimension to the political figure of the refugee. When her movement is not spatially stalled in the camp, he argues that the refugee constitutes a 'disquieting element' (2004, 127) for the sovereign nation-state because 'the so-called sacred and inalienable rights of man show themselves to lack every protection and reality at the moment in which they can no longer take the form of rights belonging to citizens of a state' (Agamben 1998, 126). As a political figure, the refugee can unfold a subversive potential which 'makes it possible to clear the way for a long-overdue renewal of categories in the service of a politics in which bare life is no longer separated and excepted' (Agamben 1998, 134). For Agamben, this subversive potential lies precisely in the refugee's status as 'a limit concept' (ibid.). Neither a political citizen of *bios* nor an individual only characterised by his human physicality, Agamben identifies the refugee's status as what he comes to describe as a form-of-life in *The Use of Bodies*. In the liminal realm between sociopolitical and biological life, it articulates 'a zone of irresponsibility, in which the identities and imputations of the juridical order are suspended' (Agamben 2016, 248).

Following Agamben, it is the crux of the biopolitical machine that it requires a liminal realm of indistinction to renew itself through its transcendence, but at the same time it thereby opens up the former as a space of withdrawal from its grasp. As a place of inoperativity where the relational reproduction of sovereign power is momentarily stalled, it can situate possible resistance to the former. But which form can this resistance in the liminal take for Agamben? In *Remnants of Auschwitz,* he draws on the *Muselmann* of the concentration camp whose life has withdrawn from all sociopolitically shaped subjectivity, but not yet passed the threshold of death, as an example for how *Homo sacer* does paradoxically both embody biopolitical governance and hold the key to overcome it. In the pure negativity which follows his 'intimate estrangement from the world' (Agamben 1999, 153), only the *Muselmann* can bear witness to Auschwitz, because he can condition 'the taking place of language as the event of a [new] subjectivity' (ibid., 164).

Desubjectified without immediate resubjectification, *Homo sacer* remains within the liminal realm spanned between *zoē* and *bios* where she can access a potentiality of epistemic possibilities unconfined by the machine of biopolitical governance. In *We Refugees,* Agamben defines the subversive potential of the refugee, present within sovereign territory without being subject to its rules, in an analogous way. He argues that the 'permanently resident mass of noncitizens, who neither can be nor want to be naturalized nor repatriated' (Agamben 2004, 117) challenges and ultimately undermines the national-sovereign foundation of the European topography and paves the way for a new Europe of urban spaces connected through 'a relationship of reciprocal extraterritoriality . . . as cities of the world' (ibid., 118). While the firm optimism of Agamben's account challenges easy dismissals of his political theory as pessimistic and disinterested in actual political change (see also Whyte 2013), I believe that Agamben's theorisation of the liminal refugee is not only problematic in its normative implications but also insufficient in its capacity to unpack the political potentiality of the recent refugee movement within the EU.

THE LIMINAL REFUGEE: RESISTANCE BEYOND DESUBJECTIFICATION, GOVERNANCE BEYOND EXCLUSION

Firstly, I argue that, even leaving aside the *Muselmann* of the concentration camp and focusing on the refugee, the assumption that he 'neither can be nor wants to' leave his liminal status and obtain citizenship or return to his home country seems rather naïve considering the living conditions which refugees face within Europe. Organised protest movements, such as the struggle for political recognition by the stateless *sans-papiers* in France in the 1990s (Nail 2015a; Doty 2011), or the recent, repeated hunger strikes of refugees for an unconditional right to asylum in Germany (Blum and Rost 2016) tell a different story. They suggest that, on the one hand, refugees seek access to the political community of the state they reside in to actively shape its governmental rules from the inside. On the other hand, the numbers of voluntary repatriates have increased significantly and now double the amount of deportations in Germany, the country which has received the highest number of asylum seekers in the wake of the 'refugee crisis' (BAMF 2016). While politically enforced and thus to be taken with more than a grain of salt, this still seems to indicate that—faced with the choice—many refugees prefer a return to their country of origin to the liminal illegality of the noncitizen.

But secondly, the revolutionary potential of the refugee as conceptualised by Agamben seems problematic in a more general sense because it is contingent on the interruption of the refugee's movement and the withdrawal of her political agency (Owens 2009, 574–79). Only a politically inoperative refugee, an empty 'form of life' stalled in the middle ground between a quasi-natural *zoē* which is surpassed and the political membership of *bios* which is not yet reached, can escape the perpetual motion of the biopolitical machine. But the refugee is never just a ready-made 'liminal concept'. The refugee movement stalled by the sovereign machine of inclusive exclusion is rather always the expression of a political will. By excluding the former from his conceptualisation of the refugee as political figure, I argue that Agamben theoretically performs a move complicit with the functionality of biopolitical governance. Agamben seeks to redefine the political desubjectivation of *Homo sacer* which he identifies as necessary part of contemporary governance towards a resource for political resistance. But by doing so, he invents his own *Homo sacer* as a political tool and empty form to be filled with a political force, which can never be his own.

But beyond these problematic implications for the political agency of the refugee, I further believe that Agamben's theoretical focus on the coupling of exception and exclusion anchored in a spatial liminality seems unsuitable to understand a number of other dimensions which form part of the European political reaction to the 'refugee crisis'. In France, new legislation taking effect at the height of the perceived crisis in November 2015 was not aimed at the exceptionalist tightening of control and the permission to use force and deter migrants. It was, rather, designed to make the existing governmental management of refugees more efficient by allocating additional personnel and housing (RFI 2015). When the German minister of the interior Thomas de Maiziere unilaterally announced withdrawal of the general recognition of Syrian migrants as refugees under the Geneva Convention to suspend the included residence permit for relatives, his exceptionalist advance was met with severe criticism and hastily taken back (Kulms 2015). Rather than awaiting a singular, sovereign decision 'on the exception' (Agamben 2005, 30), refugee movement here seems to provide an opportunity to demonstrate the sufficiency and effective functioning of the existing political-legal regime.

Even Hungary's emergency legislation was politically framed as necessary to guarantee the correct application of the EU's Dublin III Regulation according to which every asylum seeker must be registered in the European country of her arrival (BBC 2016). The political inclusion of the refugee movement through the sovereign whose realm she enters seems to be just as characteristic for the political management and the public reception of the EU 'refugee crisis' as Agamben's territorial exclusion—and illustrated by similarly powerful images (Mortensen 2016). Especially in the early stages of the perceived crisis, several EU countries emphasise that their national territories are open to arriving refugees. Attending a 'Refugees Welcome' rally in Stockholm, Swedish prime minister Stefan Lofven insists that 'Europe does not build walls' (Lofven quoted in DW 2016). The positive framing of the 'refugee crisis' as a political task which will be embraced and can be successfully managed is most explicitly expressed by the German government in the form of Merkel's repeated assurance 'Wir schaffen das'—'we can do it' (Merkel quoted in Stockrahm 2016).

The Hungarian images of violent border securitisation find their point of contrast in the citizens of Munich depicted cheering as they welcome refugees at the city's central station. Research on the media portrayal of the EU 'refugee crisis' suggests that throughout Europe, the humanitarian frame of the refugee as a victim without agency to be saved is more prominent than the exceptionalist trope that refugees constitute a security threat.[2] In the course of 2016, European governments have certainly homogenised their political responses to the 'refugee crisis' towards a general tightening of border securitisation combined with an increasingly exceptionalist rhetoric in the face of sustained right-wing criticism (Crawley 2016). Against this background, one could treat those early instances of political-legal and spatial inclusion as governmentally irrelevant and momentary ambiguities promptly resolved by the ontologico-political machine of sovereign power.

But even so, I believe that a more general point is to be made here regarding the capacity of Agamben's theory to unpack refugee movement in the different dimensions of its politicisation. The limitations of Agamben's thought to understand the governance of migration movements have certainly been pointed out before: it overlooks exactly the liberal-humanitarian dimension of refugee governance which complements violent securitisation pointed out above, overly relies on executive exceptionality which oversimplifies the way in which spaces of confinement and borders are governed, and cannot account for the political productivity of material conditions (De Genova 2013; Doty 2011; Owens 2009; Ticktin 2005). Yet, I argue, what these critiques miss out—and leave intact—is the theoretical underpinning which supports

Agamben's simplified understanding of governmental power as necessarily excluding and exceptional: the static, clearly delineated spatiality of his sovereign realm.

For Agamben, governance requires the 'kenomatic' (2005, 48) state of exception; it requires static spaces of exclusion, camps, and national borders, because it is their momentary transcendence which constitutes the political sovereign as deus ex machina. For this reason, Agamben's theory cannot explain a governance of refugees which diverts from spatial exclusion but beyond that seems unsuitable to understand the reproduction of governmental power under conditions increasingly shaped by global capital flows rather than by the sharp boundaries of national-sovereign territories. Here, Etienne Balibar suggests Deleuze and Guattari's '[g]eneralized concept of "territory"' (2009, 192) as a theoretical resource to understand the political governance of migration movements in a way which is dissolved from and always more complex than its spatial confinement.

> To 'territorialize' means to assign 'identities' for collective subjects within structures of power, and, therefore, to categorize and individualize human beings and the figure of the 'citizen' . . . is exactly a way of categorizing individuals. Such a process is possible only if other figures of the 'subject' are violently or peacefully removed, coercively, or voluntarily destroyed. (Balibar 2009, 192)

Deleuze and Guattari replace the geographical territory as the foundation of governmental power with the idea that sovereign authority can only persist through the constant, inclusive territorialisation of movement (Deleuze and Guattari 1987, 312–16). In this sense, I suggest that political-legal inclusion and exceptional exclusion should be understood as two interchangeable faces of the same phenomenon: the governmental direction of refugee movement towards the territorialisation of the political status quo. I draw on Deleuze's theory of movement-images firstly to resolve Agamben's 'bare life' from its spatially connoted liminality and instead redefine it as a political frame used to direct refugee movement. Secondly, I show how Deleuze's idea of the time-image which opens movement to durational complexity allows for connecting resistance and refugee movement in a way which incorporates rather than negates the political agency of refugees.

THE KINOPOLITICS OF THE MOVEMENT-IMAGE

In *The Figure of the Migrant,* Thomas Nail (2015b) develops a Deleuzian theory of migration as necessary condition for the persistence of a governmental apparatus reproduced through the regulation of flows: 'Flows of property that is sold, flows of money that circulates, flows of production and means of production making ready in the shadows, flows of workers becoming deterritorialized' (Deleuze and Guattari 1983, 223). Under the conditions of decoded and deterritorialised flows which global capitalism is attended by, sovereign governance can only reproduce its authority through their reactive political framing: 'It is now up to the State to recode as best it can, by means of regular or exceptional operations, the product of the decoded flows' (ibid.). When quoting Deleuze and Guattari, Agamben's focus on productive stasis means that he misses out the politically productive link between territoriality and movement which Deleuze and Guattari point to when they argue that '[s]overeignty only rules over what it is capable of interiorizing'.

Nail (2015b, 27–28) argues that the political management of migration flows between exclusion and inclusion is an example for this reproduction of a sociopolitical inside which is not static but a dynamic set of relations in constant need of reterritorialisation through governmental coding. Agamben's 'kenomatic' governance of refugee movement that emerges from a politically productive stasis is here confronted with a 'kinopolitics' of migration.

> Instead of analysing societies as primarily static, spatial or temporal, kinopolitics or social kinetics understands them primarily as 'regimes of motion'. Societies are always in motion: directing people and objects, reproducing their social conditions (periodicity), and striving to expand their territorial, political, juridical and economic power. (Nail 2015b, 24)

I suggest that the political practice of this kinopolitical directing process can be further unpacked with reference to Deleuze's theory of movement-images. Where Deleuze and Guattari (1987, 477–81) describe sovereign power as reliant on the coding of movement flows towards territorialised blocs of space-time, Deleuze unpacks this movement with reference to Henri Bergson as composed of images in the later published *Cinema 1*. For Bergson (1988), images of durational movement constitute the middle ground between matter and thought, or that of matter which can be perceived. But because matter-images are, like matter itself, always multiple, the question which arises for Bergson is 'not how perception arises, but how it is limited, since it should be the image of the whole, and is in fact reduced to the image of that which interests you' (1988, 18–21). Deleuze then argues that movement produces perception by oscillating between these limitations it directionally draws from the durational multiplicity of matter and reopening them to the former: '[M]ovement relates the objects of a closed system to open duration, and duration to the objects of the system which it forces to open up. Movement relates the objects between which it is established to the changing whole which it expresses, and vice versa' (Deleuze 1986, 13).

Moving away from Bergson's focus on individual-bodily perception, Deleuze shows how the movement-images of the cinema are territorialising 'blocs of space-time' (Deleuze 1986, 61) caught in this dual function of movement. The individual movement-images thus constitute a relational apparatus which performs the connection between a totalised whole and its parts. But it is not the spiritual automaton of individual thought but a memory consistent of the same totality of social, political, material relations which the machine of biopolitics governs. The logic which undergirds the connective functionality of the movement-image is here the ontologico-political logic of sovereign power, which 'proceeds by linking through rational divisions, projecting a model of Truth in relation to totality. The noosigns of the movement-image derive from a belief in the possibility of action and the stability of Truth' (Rodowick 1997, 11–12).

The device of the limiting closure of movement-images reproductive of the totality of memory is framing, the structuring and 'determination of a closed system' (Deleuze 1986, 12). The cinematic frame isolates particular events and actions and places them in a spatio-temporal sequence through which a narrative line of images is 'forestalled, anticipated' (ibid. 62) in the perception of the observer (the perception-image). In this sense, framing does not only determine the perceived content of and relation between images. It can also present them as the causal effect of particular actions—when all movement is linked to the action-image built up through its framed sequencing (ibid., 65). While the establishment of action-images dominated cinema before the Second World War, Deleuze shows how it first undergoes a shift from global to local, elliptical representation, from structural stability to a focus on events, before experiencing a general crisis when faced with a new cinematic form which uses image montages to access the openness of uncompartmentalised, durational movement (Rodowick 1997, 74–78).

I would like to draw attention to the interesting parallel between the gradual dissolution of the movement-image as dominating cinematographic framing 'in the disjointed post-war world' where the 'unities of situation and action can no longer be maintained' (Tomlinson and Galeta 1989, xv) and Deleuze and Guattari's analysis of the decline of sovereign authority caused by the capitalist decoding and complexification of flows. Developing this parallel, I use Deleuze's theory of movement-images as a resource for a more philosophically grounded alternative to the analysis of 'framing' already established in media and communication research (Entman 1993), which can explore the politically productive role of movement. I argue that the political framing of movement-images can be understood as one way in which the sovereign coding of flows towards reterritorialisation takes place in practice. While sovereign governance does not have the power to produce or terminate movement-flows, it can frame the condensed images through which this movement is received by and affects public perception.

Determining what is shown, from which angle, and in which sequence, sovereign framing codes movement towards the reproduction of its political authority, the action-image in which every effect is causally linked to the sovereign decision. At the same time, sovereign framing closes off the politically internalised movement from alternative codes which would distort the perception of effective political regulation. Against this background, I propose, in a slight variation of Nail's argument, to unpack the sovereign coding which precedes political reterritorialisation as the framing of movement-images towards the action-image of effective political governance. This framing can include securitisation and exceptional exclusion, but sovereign space here loses its decisive status. The movement-image is not a space but a 'spatialization of time' (Rodowick 1997, 79) which is always momentary, and always integrated within a relational flux politically directed towards the reproduction of the socioeconomic status quo. In this sense, it is not necessary to prevent refugees from entering sovereign territory if their movement can be directed in such a way that the perception it produces is causally linked to the sovereign action-image and reproduces the truth and effectiveness of its political authority.

With Deleuze, the bare life of *Homo sacer* can be dissolved from its spatial situatedness and instead be identified as such a political frame, expressed in the discursively dominant distinction between 'refugee' or 'asylum seeker' and 'economic migrant'. Those who escape violence and prosecution devoid of resources and political agency, with nothing left but their bare lives, can be allowed to enter the country. Politically neutralised through the sovereign frame of bare life, their movement is, regardless of its situatedness relative to the realm of sovereign governance, already directed towards its reterritorialisation. Excluded, on the contrary, are 'economic migrants' who don't fit the frame of *Homo sacer*. Their movement is the intentional expression of a political agent distinct from the territorial refrain and thus potentially jeopardises the politically reproductive synchronisation of movement. This understanding of *Homo sacer* as a flexibly applicable political frame makes it possible to include the dimension of humanitarian control which Miriam Ticktin—albeit within Agamben's spatially static framework—diagnoses as missing from the framework of sovereign exceptionalism and which 'privileges forms of life or humanity not constituted as rights-bearing individuals, but as corporeal victims of sexual violence, innocent, nonagentive, and apolitical' (2005, 367).

In the case of the European 'refugee crisis', the refugee who fits the frame of *Homo sacer*, it seems, is Syrian. The initially welcoming governmental approaches as well as the humanitarian public framing and civil action are predominantly focused on Syrian refugees, whose political identities as *homines sacri* are well documented in the mediatised images of the Syrian Civil War (EC 2016a). In Germany, the arrival of thousands of migrants whose asylum claims need to be registered led to the adoption of new laws, which allowed local authorities to quickly deport migrants from Balkan countries. This procedure of rejecting asylum claims solely based on nationality has been criticised as contingent and unjust, in this case particularly against the background of the systematic violation of minority rights within numerous Balkan states (Gathmann and Maxwill 2015). Interesting at this juncture is also how the United Kingdom, faced with increasing European pressure, agreed to receive twenty thousand Syrian refugees from UN camps erected in Jordan in 2015. On the contrary, the five thousand people living in the Calais 'jungle camp' waiting for their chance to cross the border were not allowed to enter the country. While Syrian nationals also form one of the largest ethnic groups represented here (BBC 2016), the Calais refugees did not passively await saving through the UK government; they demanded access to its sovereign realm.

Because their self-directed movement cannot readily be made useful to reproduce the action-image of political control, it is kept separate from the relational whole of the sovereign realm. In precisely this sense, conservative Dutch prime minister Mark Rutte demands that refugees 'behave normally, or go away' (Rutte quoted in Osborne 2017)—that they do not resist the synchronisation of their expressions towards the reproduction of the sociopolitical status quo or face exclusion. Turning to the right-wing criticism of the reception of refugees, it certainly aims at the narrowing, the zooming-in of the political frame of bare life in combination with a strengthening of the framing of refugees as security threats and ethnic-religious others. But interestingly, those other frames justify the necessity to exclude refugees precisely with the presupposed political agency and culturally shaped subjectivity of the refugee—with those characteristics that exceed her status as *Homo sacer*—strengthening the governmental frame of bare life at its margins. For instance, right-wing propaganda websites such as Breitbart or politically incorrect express particular outrage over the idea that refugees should be provided with a degree of financial independence and resources beyond the goods they need to immediately sustain their bodily life (Williams 2016; politically incorrect 2015).

OPENING UP THE FRAME: RESISTANCE IN THE TIME-IMAGE OF REFUGEE MOVEMENT

Deleuze's theory helps to complicate Agamben's ultimately static, spatially grounded sovereign apparatus and the way it captures refugee movement.

Rather than including migration only negatively as an instance for exceptional control which must, however, be contained geographically to keep the inside/outside machine of biopower in motion, the Deleuzian approach I suggest provides an understanding for how sovereign power can include movement to reproduce sovereign power. Detached from its spatial connotations, bare life becomes a political frame which neutralises the potentially disruptive force of refugee movement and directs it towards the reterritorialisation of the political status quo. As a political frame which internalises and homogenises refugee movement, bare life, like the shot of a film camera, is variable in how narrow or wide the movement is which it catches, and in its position compared to other frames. However, importantly, all frames, even if focused on the securitisation and expulsion of refugees, are only activated by the territorial inclusion of their movement. In contrast to Agamben's static apparatus of capture, a sovereign power which reproduces itself through the framing of movement is not threatened by the former but requires its territorial presence to make it politically productive.

In the light of the extreme and extremely polarised politicisation of EU migration and its consequences for the European political landscape, this Deleuzian perspective suggests that, in fact, both exclusionary-exceptional and inclusive frames form two sides of the same old coin, which is self-reproductive governmental control. What is reproduced is the general responsivity of governance, the action-image focused on the possibility of effective political management, through a sovereign frame which connects introversive movement to the duration of the political territory, and for which bare life is only one example. This undermines Agamben's rather naïvely optimistic appraisal of the mere presence of refugees on sovereign territory as politically subversive. With Deleuze, the liminality of the refugee cannot be thought as an empty, immanent form-of-life which can give way to a newly emergent subjectivity, because withdrawal from the relational shaping of the onto-political machine is illusionary. In Deleuze, immanence is the durational whole of all movement-relations, which opens up closed territorial systems towards change but is always itself formed by previous movements of territorialisation (Deleuze 1986, 59–60).

However, I argue that Deleuze's Bergsonian conceptualisation of every singular movement-image as secondary to a temporal duration opens up the possibility to think resistance from within the sovereign frame in a way that does not rely on the annihilation of the refugees' political agency. Through Bergson, Deleuze reverses the tradition of Einsteinian physics which gives priority to the dimension of space when understanding movement, which, as we have seen, comes to the fore in Agamben. Instead he identifies movement as dependent on the relations of time (Clarke 2002, 168–70). Resistance, in this sense, must not do away with the closing framing of movement-images but challenge it through the constant reopening of movement to the duration as a whole (Deleuze 1989, 28–29). Resistance must be thought as an aberrant movement. Not unlike the both conditioned and genuinely creative agency which Foucault's (1988) aesthetics of existence aims at, it dissolves people and things from their territorial situatedness and connects them to the whole of matter-images in the time-image of duration (Deleuze 1989, 39–42). The time-image, or crystal of time, due to its multifaceted and ever-growing nature, opens the closed system of the frame towards the totality of a scene.

It does not annihilate the framed roles but challenges them through the making-present of a multiplicity of alternative relations in 'the whole of the real, life in its entirety, which has become spectacle' (Deleuze 1989, 84).[3] The time-image of movement can never be perceived as such, because it is a pure durational complexity in which all possible pasts, presents and futures coincide. Every perception-image which is entailed and can be actualised from its crystalline multiplicity carves out a particular temporal lineage, a particular relation of past, present and future (Bergson 1988, 144–46). But every evental movement which reopens the multiplicity of the time-image always includes the possibility to actualise a new past-future lineage and cut out a different frame from the scene of the whole of matter-images (Deleuze 1989, 109–13). With regard to the cinematographic frame, Deleuze suggests montage as a strategy of resistance from the inside that subverts the closed system of framed movement-images. Through the 'juxtaposition of images, or irrational "cuts" between scenes, or even by the dissonance between sounds and visual imagery' (Clarke 2002, 172–73), montage relationally produces a time-image; it reveals how multiple 'sheets or strata of past coexist' (Deleuze 1989, 116) in their relation to the centre which is the evental present.

Transferred to the political framing of refugee movement as a means of territorialisation, this means that resistance emerges from within the public, mediatised relation of movement-images but connects or disjoints these in unexpected ways to produce a crack in the sovereign frame, which releases relational ambiguities. In the time-image opened up in this sense, alternative temporal lineages, spatial connections and representations become available for actualisation—historical connections to examples for the successful political integration of refugees or lines which link refugee movement to global economic power structures, past political decision and the ad hoc political agendas within Western democracies. It must be pointed out that a time-image of EU refugee movement certainly makes available lineages for actualisation which not only conflict with democratic governance in its ossified sovereign form but also the value consensus it founds on. The political potential of the time-image lies in its relational multiplicity, but the actualisation of the relations it makes accessible must always be politically made, and is thus always political.

With Bergson, Deleuze defines these relational connections made available in the duration of time through movement as the nature of subjectivity. '[T]he only subjectivity is time, non-chronological time grasped in its foundation, and it is we who are internal to time, not the other way round. That we are in time looks like a commonplace, yet it is the highest paradox. Time is not the interior in us, but just the opposite, the interiority in which we are, in which we move, live and change' (Deleuze 1989, 82). In this sense, Deleuze's breaking open of the movement-image in the crystal of time finally also resolves the problematic status of subjectivity in Agamben. While acknowledging the politically constituted nature of all individual subjectivity, Deleuze does not abandon the possibility of resistant political agency. His time-image is not a forced withdrawal from all sociopolitical relations to a liminal less-than which seems at best ambiguous it is political productivity. It is rather the political push towards a more-than, a making-productive of the subjective multiplicity of these relational 'durations which are inferior and superior to man, all coexisting' (Deleuze 1989, 118). These do include the identities, ideas and political interests of the refugees thereby freed from the confinement of their sovereign framing. While neither normatively essentialising these nor disregarding their political conditionality, resistance emerges from the crack in the political frame of *Homo sacer* as the 'good' refugee who must be saved because she comes with nothing but her bare life.

NOTES

I would like to thank Iain MacKenzie and Shubranshu Mishra for their insightful and helpful comments on this chapter.

1. While both 'refugee crisis' and 'migrant/migration crisis' are used interchangeably to refer to recent migration into the European Union (Berry et al. 2015, 13–14), this chapter employs the term 'refugee crisis'.
2. While 'humanitarian themes were more numerous in Italian coverage (50.6%) than in Swedish (47.1%), British (37.3%), German (37.1%) or Spanish reporting (32.5%)' (Berry et al. 2015, 8) they were significant for all the analysed countries. In comparison, refugees were discussed as security threats 'in 10.1% of articles in Italy, 9.2% in Spain, 8.5% in Britain, 4.8% in Germany and 2.3% in Sweden' (ibid.).
3. Especially with regard to photography, Agamben also considers images as a resource to escape sovereign capture (Noys 2009). However, contrary to Deleuze, the political potentiality of the image here relies on its static quality, the freezing of sociopolitical relations it performs and thus again falls victims to the political and theoretical issues identified with regard to a resistance which necessitates withdrawal to a liminal space here.

BIBLIOGRAPHY

Agamben, Giorgio. 1998. *Homo Sacer: Sovereign Power and Bare Life*. Palo Alto, CA: Stanford University Press.

Agamben, Giorgio. 1999. *Remnants of Auschwitz: The Witness and the Archive*. Palo Alto, CA: Stanford University Press.

Agamben, Giorgio. 2004. "We Refugees." *Symposium Summer 1995* 49 (2): 114–19.

Agamben, Giorgio. 2005. *State of Exception*. Chicago: University of Chicago Press.

Agamben, Giorgio. 2011. *The Kingdom and the Glory*. Palo Alto, CA: Stanford University Press.

Agamben, Giorgio. 2016. *The Use of Bodies: Homo Sacer IV, 2*. Palo Alto, CA: Stanford University Press.

Alcalde, Xavier. 2016. "Why the Refugee Crisis Is Not a Refugee Crisis." *Peace in Progress* 29: 1–10.

Aradau, Claudia, and Rens van Munster. 2011. *Politics of Catastrophe: Genealogies of the Unknown*. London: Routledge.

Balibar, Etienne. 2009. "Europe as Borderland." *Society and Space* 29: 190–215.

BAMF. 2016. "Freiwillige Rückkehr mit REAG/GARP." *Bundesamt für Migrationsforschung*. http://www.bamf.de/DE/Infothek/Statistiken/FreiwilligeRueckkehr/freiwillige-rueckkehr-node.html. Accessed 31 January 2017.

BBC. 2016, March 4. "How Is the Migrant Crisis Dividing EU Countries?" http://www.bbc.co.uk/news/world-europe-34278886. Accessed 31 January 2017.

Bergson, Henri. 1988. *Matter and Memory*. New York: Zone Books.

Berry, Mike, Inaki Garcia-Blanco and Kerry Moore. 2015. "Press Coverage of the Refugee and Migrant Crisis in the EU: A Content Analysis of Five European Countries. Report Prepared for the United Nations High Commission for Refugees." http://www.unhcr.org/56bb369c9.html. Accessed 31 January 2017.

Blum, Katharina, and Christian Rost. 2016, November 1. "Münchner Flüchtlinge treten in Hungerstreik." *Süddeutsche Zeitung*. http://www.sueddeutsche.de/muenchen/protest-muenchner-fluechtlinge-treten-in-hungerstreik-1.3230262. Accessed 31 January 2017.

Clarke, Melissa. 2002. "The Space-Time Image: The Case of Bergson, Deleuze and Memento." *Journal of Speculative Philosophy* 16 (3): 167–81.

Crawley, Heaven. 2016. "Managing the Unmanageable? Understanding Europe's Response to the Migration 'Crisis.'" *Human Geography* 9 (2): 13–23.

De Genova, Nicholas. 2013. "Spectacles of Migrant 'Illegality': The Scene of Exclusion, the Obscene of Inclusion." *Ethnic and Racial Studies* 36 (7): 1180–98.

Deleuze, Gilles. 1986. *Cinema 1: The Movement-Image*. Minneapolis: University of Minnesota Press.

Deleuze, Gilles. 1989. *Cinema 2: The Time-Image*. Minneapolis: University of Minnesota Press.

Deleuze, Gilles, and Félix Guattari. 1983. *Anti-Oedipus: Capitalism and Schizophrenia*. Minneapolis: University of Minnesota Press.

Deleuze, Gilles, and Félix Guattari. 1987. *A Thousand Plateaus: Capitalism and Schizophrenia*. Minneapolis: University of Minnesota Press.

Diken, Bülent. 2004. "From Refugee Camps to Gated Communities: Biopolitics and the End of the City." *Citizenship Studies* 8 (1): 83–106.

Doty, Roxanne Lynn. 2011. "Bare Life: Border-crossing Deaths and Spaces of Moral Alibi." *Environment and Planning D: Society and Space* 29: 599–612.

DW. 2016, June 21. "Sweden Approves Stricter Asylum Laws." http://www.dw.com/en/sweden-approves-stricter-asylum-laws/a-19346623. Accessed 31 January 2017.

EC. 2016a. "Syria Crisis." *European Commission Humanitarian Aid and Civil Protection Factsheet.* http://ec.europa.eu/echo/files/aid/countries/factsheets/syria_en.pdf. Accessed 31 January 2017.

EC. 2016b. "The EU and the Refugee Crisis." *European Commission Factsheet.* http://publications.europa.eu/webpub/com/factsheets/refugee-crisis/en/. Accessed 31 January 2017.

Entman, Robert. M. 1993. "Framing: Toward Clarification of a Fractured Paradigm." *Journal of Communication* 43 (4): 51–58.

Eurostat. 2017. "Asylum Statistics." http://ec.europa.eu/eurostat/statistics-explained/index.php/Asylum_statistics. Accessed 11 July 2017.

Foucault, Michel. 1988. "An Aesthetics of Existence." In *Politics, Philosophy, Culture: Interviews and Other Writings, 1977–1984*, edited by Lawrence Kritzman, 47–53. London: Routledge.

Gathmann, Florian, and Peter Maxwill. 2015, June 23. "Balkan-Flüchtlinge—worum es in dem Streit geht." *Spiegel Online.* http://www.spiegel.de/politik/deutschland/fluechtlinge-vom-balkan-warum-die-politik-darueber-streitet-a-1044966.html. Accessed 31 January 2017.

Goddeeris, Idesbald. 2016. "Historical Reflections on the 2015 Refugee Crisis." *Global Kring voor Internationale Betrekkingen* 20 (1): 24–27.

Holehouse, Matthew, and Isabelle Fraser. 2015, November 13. "Migrant Crisis: European Council President Tusk Warns Schengen on Brink of Collapse." *Telegraph.* http://www.telegraph.co.uk/news/worldnews/europe/eu/11991098/Migrant-crisis-Donald-Tusk-warns-that-Schengen-is-on-brink-of-collapse-latest-news.html. Accessed 31 January 2017.

Kennedy, Dane. 2016. "Is the European Refugee Crisis Unprecedented? Symposium at the German Historical Institute Provides Historical Perspective." *American Historical Association Today.* http://blog.historians.org/2016/04/the-european-refugee-crisis-in-historical-perspective/. Accessed 31 January 2017.

Kmak, Magdalena. 2015. "Between Citizen and Bogus Asylum Seeker: Management of Migration in the EU through the Technology of Morality." *Social Identities* 21 (4): 395–409.

Kulms, Johannes. 2015, November 10. "Familiennachzug von Flüchtlingen Kritik und Rückendeckung für de Maizière." *Deutschlandfunk.* http://www.deutschlandfunk.de/familiennachzug-von-fluechtlingen-kritik-und-rueckendeckung.1766.de.html?dram:article_id=336457. Accessed 31 January 2017.

Martin, James. 2015, April 20. "Katie Hopkins Wrote This in the Sun about Migrants and Now Everyone Is Really Angry." *Huffington Post.* http://www.huffingtonpost.co.uk/2015/04/18/katie-hopkins-russell-bra_n_7091674.html. Accessed 31 January 2017.

Mortensen, Mette. 2016. "'The Image Speaks for Itself'—or Does It? Instant News Icons, Impromptu Publics, and the 2015 European 'Refugee Crisis.'" *Communication and the Public* 1 (4): 409–22.

Nail, Thomas. 2015a. "Alain Badiou and the Sans-Papiers." *Angelaki* 20 (4): 109–30.

Nail, Thomas. 2015b. *The Figure of the Migrant*. Palo Alto, CA: Stanford University Press.

Noys, Benjamin. 2009. "Separation and Reversibility: Agamben on the Image." *Filozofski Vestnik* 30 (1): 143–59.

Osborne, Samuel. 2017, January 24. "'Behave Normally, or Go Away' Dutch PM Warns Immigrants Weeks before Vote." *Independent*. http://www.independent.co.uk/news/world/europe/dutch-prime-minister-mark-rutte-immigrants-refugees-netherlands-behave-normally-go-away-geert-a7543196.html. Accessed 31 January 2017.

Owens, Patricia. 2009. "Reclaiming 'Bare Life'?: Against Agamben on Refugees." *International Relations* 23 (4): 567–82.

Politically incorrect. 2015, October 31. "Gutschein statt Bargeld für 'Flüchtlinge'." http://www.pi-news.net/2015/10/p488880/. Accessed 31 January 2017.

RFI. 2015, November 11. "French Reforms to National Asylum System Take Effect." http://en.rfi.fr/france/20151102-french-reforms-national-asylum-system-take-effect. Accessed 31 January 2017.

Rodowick, David N. 1997. *Gilles Deleuze's Time Machine*. Durham, NC: Duke University Press.

Rygiel, Kim. 2011. "Bordering Solidarities: Migrant Activism and the Politics of Movement and Camps at Calais." *Citizenship Studies* 15 (1): 1–19.

Spijkerboer, Thomas. 2016. "Minimalist Recollections on Europe, Refugees and Law." *European Papers* 1 (2): 533–88.

Spinder, William. 2015. "2015: The Year of Europe's Refugee Crisis." *UNCHR Tracks*. http://tracks.unhcr.org/2015/12/2015-the-year-of-europes-refugee-crisis/. Accessed 31 January 2017.

Stockrahm, Sven. 2016, July 30. "Wir schaffen das nicht?" *Zeit Online*. http://www.zeit.de/politik/deutschland/2016-07/fluechtlingskrise-deutschland-angst-terror-radikalisierung. Accessed 31 January 2017.

Ticktin, Miriam. 2005. "Policing and Humanitarianism in France: Immigration and the Turn to Law as State of Exception." *Interventions* 7 (3): 346–68.

Tomlinson, Hugh, and Robert Galeta. 1989. "Translators' Introduction." In *Cinema 2: The Time-Image*, edited by Gilles Deleuze, xv–xviii. Minneapolis: University of Minnesota Press.

Vaughan-Williams, Nick. 2015. *Europe's Border Crisis: Biopolitical Security and Beyond*. Oxford: Oxford University Press.

Williams, Thomas D. 2016, October 28. "European Union Offers a Million Debit Cards for Refugees in Syria." *Breitbart*. http://www.breitbart.com/london/2016/10/28/european-union-offers-million-debit-cards-syrian-refugees-turkey/. Accessed 31 January 2017.

Whyte, Jessica. 2013. *Catastrophe and Redemption: The Political Thought of Giorgio Agamben*. New York: SUNY Press.

Chapter 5

The Biopolitical Economy of 'Guest' Worker Programmes

Greg Bird

> [I]t makes no sense to pay whities to stay home while we bring in brown people to work in these jobs.
>
> —Mazer and Baker (2015)

In the literature on biopolitics a debate has emerged regarding the precise nature of the biopolitical division of life. How is the relationship between worthy, politically included and legally recognised life (*bíos* or proper life) and unworthy, politically excluded and legally unrecognised bare life (*zoē* or improper life) formulated? How is this division of life materially articulated and sustained? What can be done to disrupt it and, if possible, liberate life from this biopolitical articulation? Theorists such as Giorgio Agamben, Roberto Esposito, Michael Hardt, Antonio Negri and Maurizio Lazzarato have argued that the asymmetrical relationship between proper and improper life is far more complicated than the simple normative distinction between privileged and degraded forms of life.

The elevation of one group is not simply based on the deprecation of another—as we find in simplistic politics of recognition reduced to normative criteria—but is more deeply rooted in relationships of domination and control that are materially embedded in political and economic structures. In this latest stage of capitalism, life is appropriated, divided up, and asymmetrically distributed in what must be called the global biopolitical economy of life. Proper life is not only distinguished from improper life, the normative rationale for its domination over improper life, because its dominion is nourished and preserved by apparatuses that enable it to exploit and appropriate the life of improper lives.[1] Put in different terms, the political dimension of domination is thoroughly entangled with the economic dimension of exploitation in what can only be called a biopolitical economy that now operates on a global scale.

In this chapter, I use Canada's Temporary Foreign Worker Program (herein TFWP) as a case study to exemplify how life is expropriated and asymmetrically distributed in the global biopolitical economy. Canada's TFWP is a biopolitical apparatus that preserves, protects and ultimately enhances the lives of Canadian citizens by devaluing, disposing and appropriating the lives of migrant workers.[2] I begin with a brief demographic sketch of this programme. Although most low-skilled migrant workers are employed in occupations that are vital to the maintenance and reproduction of the life of the nation (food services, agriculture, care work, etc.), this apparatus devitalises them. It has been designed to exploit and reproduce the global division of life between those who are reduced to mere 'biological labourers' and those elevated to the status of 'vital citizens'. Devitalised, the improper lives of the labourers are used up and exploited in order to enhance the vitality of the lives of citizens. In the second part, I begin by examining how this division of life is woven through structural and common-sense racism, which is articulated in a heteronormative and essentialist notion of biological sex differences.

Then I demonstrate how devitalising the lives of migrant workers serves the dual purpose of immunising the nation and the state. First, the nation is immunised from being overexposed to potentially threatening numbers of migrant workers, whose numbers are restricted and who are subjected to a series of measures that have the cumulative effect of sterilising them while on Canadian soil. Through an examination of a series of well-documented practices, I argue that the net result is a sophisticated eugenics programme. Second, because migrant workers are categorised as nonvital lives, the state is legally immunised from bearing responsibility for anything other than their basic survival. Here, I briefly examine their relationship to health care, the nature of the work contracts and the practice of deportation.

THE DEMOGRAPHICS OF CANADA'S TEMPORARY FOREIGN WORKER PROGRAM

> The fact that the growing flows of immigrants are thought (entirely erroneously) to be one of the worst dangers for our societies also suggests how central the immunitary question is becoming. Everywhere we look, new walls, new blockades, and new dividing lines are erected against something that threatens, or at least seems to, our biological, social, and environmental identity.
>
> —Esposito (2013a, 59)

Canada has a long history of importing migrant, sometimes indentured, labourers, which dates back to the nineteenth century.[3] Some of these programmes were products of bilateral and multilateral trade agreements, others were part of a more comprehensive package of policies aimed at recruiting workers, while others were part of a 'Home Children' scheme that deported around 100,000 children from Scotland and England who worked in Canada as indentured farm and domestic labourers (1860s–1930s).[4] The father of our currently popular prime minister Justin Trudeau, Pierre Trudeau, established the first large-scale, fully operational migrant labour programme called the Non-Immigrant Employment Authorization Program (herein NIEAP) in 1973. Trudeau senior was not only a charismatic, sexy, modern and suave figurehead, like his son, but his period of political rule (1968–1979 and 1980–1984) has also been memorialised as a period of great modernisation and social progress in the nationalist narrative of Canada. Trudeau senior has been credited with overseeing a fundamental transformation that turned a formerly backwards, conservative and white British settler colony into a world-leading liberal democracy that values human rights and multiculturalism. Without entering into how contentious each of these claims is, what needs to be noted is that Trudeau senior's liberal government established the NIEAP while it was simultaneously engaging in a rebranding campaign to promote its new official doctrine of 'multiculturalism'.[5]

To the rest of the world, Canada became a beacon of modern liberal democracy, an open and inclusive society with culturally diverse nations, but below the surface of this official rhetoric a more sophisticated system of discrimination and exploitation was established along racialised, gendered, classed, abelist and heteronormative lines. It is only after considering this other dimension of Canadian immigration policies, in combination with the now centuries-long process of genocide against indigenous peoples of Canada (first biological, second cultural) by white settler colonialism, that the international brand projected by successive Canadian governments as the model state with the best practices in developing an inclusive and diverse society, the 'multicultural safe haven' or 'Canada the Good', has to be seriously questioned. Like all marketing campaigns, this branding exercise is strategically selective of what is included and excluded. Differential inclusion is central to the 'multicultural mosaic', which remains a white settler colonial state that has yet to seriously engage in a genuine reconciliation process and continues to discriminate and exploit migrant workers.

On the immigration side of this history, the Immigration Act of 1976 was supposed to open up Canada's borders by moving away from its previous practices of discriminating against specific groups of people based on ethnicity, national origin or race. A point system was introduced that would determine who would be qualified on the basis of human and economic capital. The new 'prohibited classes' were those that could become a 'burden on social welfare' and/or 'health services'. In practice, however, this points system introduced a whole new array of discriminatory criteria. In addition to a person's class and abilities, the points system effectively blocked entry for the vast majority of people from the Global South. At the same time, the NIEAP and its successor TFWP were established to catch and provide a second, albeit temporary, entry point for the same groups of people who could not qualify for citizenship.[6]

Since the 1970s, the Canadian government has increasingly turned to migrant workers to supply its labour market with a much more manageable, flexible and exploitable pool of cheap labour. For example, when the NIEAP was founded in 1973, 57 percent of people entering Canada who were classified as 'workers' held the status of being a 'permanent resident', but by 1993 this percentage dropped to 30 (Sharma 2001, 424). By 2012, 75 percent of workers entering Canada were migrant workers (CIC 2014). In a systematic study of this shift, Judy Fudge and Fiona MacPhail point out that between 1983 and 2007, there has been a 'dramatic increase in the number of temporary foreign workers from 97,500 to 302,300' (2009, 17). The most recent figures available show that in 2012 the number of temporary foreign workers in Canada rose again up to 491,547 (see Table 5.1).

This pattern accelerated under Stephen Harper's Conservative Party of Canada's near decade in power (2006–2015). Between 2005 and 2012, the number of temporary migrant workers increased by 219.4 percent (17.1 percent annually); yet there was a slight decline in permanent residents. In fact, the same year the Conservatives took power, Canada accepted slightly more temporary migrant workers than permanent residents (101.5 percent). In 2012, this ratio was nearly doubled at 190.6 percent. If we limit the numbers to the economic categories, the increase is even more staggering. By 2012 there were 491,547 temporary migrant workers but only 160,819 permanent residents in the economic class, which is a ratio of 305.7 percent. These figures clearly indicate that temporary migrant workers became the preferred source for immediately supplying the Canadian labour market with new workers (see Table 5.1).[7]

There are several obvious demographic dimensions to this new pattern. First, there is a class-based distinction between low-skilled and high-skilled workers. Low-skilled workers are filtered into the temporary foreign worker category, whereas high-skilled workers are placed in the immigration track. For example, of the 77,937 permanent residents in 2012 who had an intention to work and an identifiable occupational skill level, the vast majority (83.7 percent or 65,250) fell under high-skilled National Occupational Classifications.[8] These high-skilled workers and their families are highly sought after and thus granted permanent residency. To solidify this distinction, the Conservative government introduced a new programme called the Canadian Experience Class in 2008, which established a fast-track system

Table 5.1 Number of Temporary Foreign Workers and Permanent Residents in Canada (1987–2014)[a]

Temp Class		*Permanent Resident Classes*					*Temp/Perm %*	
Date	*Foreign Workers*	*Family*	*Economic*	*Refugee*	*Other*[b]	*Total*	*TFW / Perm Eco Class*	*TFW / Total Perm Class*
1987	131,752	53,799	74,077	21,455	2,748	152,079	177.9%	86.6%
1992	171,737	101,109	95,787	52,337	5,567	254,800	179.3%	67.4%
1997	147,679	59,925	128,349	24,307	3,454	216,035	115.1%	68.4%
2002	181,794	62,288	137,863	25,111	3,786	229,048	131.9%	79.4%
2005	224,022	63,373	156,313	35,775	6,781	262,242	143.3%	85.4%
2006[c]	255,331	70,515	138,248	32,499	10,378	251,640	184.7%	101.5%
2007	299,830	66,240	131,244	27,954	11,315	236,753	228.5%	126.7%
2008	361,564	65,581	149,068	21,859	10,739	247,247	242.5%	146.2%
2009	403,232	65,207	153,491	22,850	10,624	252,172	262.7%	159.9%
2010	431,326	60,223	186,916	24,697	8,853	280,689	230.8%	153.7%
2011	445,587	56,449	156,117	27,873	8,309	248,748	285.4%	179.1%
2012	491,547	65,008	160,819	23,094	8,966	257,887	305.7%	190.6%
2013	N.A.[d]	81,843	148,155	23,831	5,194	259,023	N.A.	N.A.
2014	N.A.	66,661	165,089	23,286	5,368	260,404	N.A.	N.A.

Source: Compiled from Citizenship and Immigration Canada, Facts and Figures 2011, 2012 and 2014, Immigration Overview: Permanent and Temporary Residents.

[a]This table is an updated and altered version of Table 1 in Fudge and MacPhail 2009. Some of my figures are slightly different because I have used the 2011, 2012 and 2014 reports, while Fudge and MacPhail used the 2008 report. The column for the temporary class includes the categories 'initial entry', 're-entry', 'still present', which accounts for the total entries and those still present.

[b]I have combined two categories in this column, which Statistics Canada treats as separate categories: 'category not stated' and 'gender not stated'. They do not contribute substantially to the overall total.

[c]Stephen Harper's Conservative Party of Canada became the official governing party when they won the federal election on January 23, 2006.

[d]Stats Canada no longer publishes detailed accounts of the total number of temporary residents working in Canada.

to enable these workers to enter into the permanent pool of residents (Fudge and MacPhail 2009, 15). Participants were encouraged to bring their family members with them, which gives them more incentive to apply for permanent settlement. Then in 2014, the same government split the Temporary Foreign Worker Program into two. The high-skilled workers would be streamed into the new International Mobility Programs (IMPs), now under the administration of Citizenship and Immigration Canada, which grant 'greater mobility' to high-skilled workers to give them more incentives to permanently settle in Canada (ESDC 2014). Low-skilled workers, on the other hand, remain in the TFWP administered by Employment and Social Development Canada. This new administrative division of labour clearly signals that low-skilled workers are to be treated as mere economic entities without the prospect of obtaining citizenship. Like most guest worker programmes, the government and employers have promoted this split as a necessary measure to find workers for jobs that Canadian citizens are no longer willing to perform.

Activists and scholars have pointed out that the class-based skill criteria employed in these programmes are also used to distinguish between workers from the Global South and the Global North. In 2005, over two-thirds of those in the high-skilled pool were from Europe and the United States; however, the majority of those from Asia and the Pacific (59 percent) and the vast majority of workers (85 percent) from the Americas (excluding the United States) were placed in the low-skilled pool (Fudge and MacPhail 2009, 21). In 2015, the top six source countries for new temporary migrant workers who were actually tracked, less than a third of the workers in the programme, were from Mexico (30,518), Jamaica (11,860), the Philippines (8,368), the United States (7,283), Guatemala (6,400) and India (3,266) (ESDC 2016). With the recent distinction between two streams, this division has been further exacerbated. The government made this clear on its primer 'Improving Clarity, Transparency and Accountability of the Temporary Foreign Worker Program' by stating that the main source countries for the revised low-skilled TFWP are 'developing countries', whereas those for the new high-skilled IMPs are 'highly developed' countries (ESDC 2014).

BIOPOLITICAL DIMENSIONS OF CANADA'S TEMPORARY FOREIGN WORKER PROGRAM

Biopolitical analyses of immigration policies, border control mechanisms, and the regulation of international migration provide detailed and nuanced accounts of the measures that have been taken to surveil, manage, regulate and control populations. For example, Nikolas Rose claims, 'biological presuppositions, explicitly or implicitly, have underlain many citizenship projects, shaped conceptions of what it means to be a citizen, and underpinned distinctions between actual, potential, troublesome, and impossible citizens' (2007, 132). At the heart of these analyses rests Foucault's famous analysis of 'state racism' (2003) and his examination of the 'techniques' that subjugate bodies and control populations (1981). Leerom Medovoi argues that in our contemporary era 'the regulation of the life of the population is itself conceived on the model of war', specifically a 'race war' (2007, 53). Another prominent scholar in this area, Didier Bigo, argues that there has been a drastic shift in the relationship between the state and the population. The state is no longer concerned with the welfare of its citizens so much as with security. One of the primary sites of this transformation is the 'securitization of immigration', where states play on the fears of their citizens by 'designating potentially dangerous minorities' in order to create 'ever more restrictive norms' and strengthen the 'internal security state' (Bigo 2002, 81–82). In one of the first biopolitical analyses of Canada's TFWP, J. Adam Perry argues that since 'the primary function of the state towards its population is to preserve life', he continues, 'then it must be possible for the state to be granted the power to conduct violence against those that could put the health of the collective body at risk' (2012, 197). What is missing in these analyses, however, is a more substantial examination of the ways that the political and the economic have become thoroughly integrated in the biopolitical economy of life.

Esposito's theory of immunisation offers a slightly new perspective on how the collective body can be immunised from potential pathogens while simultaneously benefitting from their presence within its borders. Immunisation represents a combination of law and biology. First, in biomedical terms, to vaccinate a patient 'against a disease', Esposito argues, 'you have to introduce a controlled and tolerable portion of it into the organism' (2013a, 61). Whether the patient is an individual or a political body, immunisation functions by 'introducing within it a fragment of the same pathogen from which it wants to protect itself, by blocking and contradicting natural development' (2008, 46). What is novel about this process is that the aim is no longer to negate and destroy potential diseases that put our lives in danger; rather, immunisation works by 'outflanking and neutralizing' its effects (2011, 8). 'Evil', he continues, 'must be thwarted, but not by keeping it at a distance from one's borders; rather, it is included inside them. The dialectical figure that thus emerges is that of exclusionary inclusion or exclusion by inclusion' (ibid.). Brett Levinson notes that although biomedical immunisation works by introducing 'a little bit of illness to prevent wider spread', it does not stop the 'production' of the illness, just its 'reproduction' (2010, 244). Put in terms of migration, governments have but two choices: accept 'illegal' migration or find a legal means to regulate, manage and ultimately limit it.

Second, from a political-legal perspective, immunity frees a person from the duty and obligation to perform services on behalf of others. An immunised person is exempt 'from the obligation of the *munus*, be it personal, fiscal, or civil' (Esposito 2011, 5).[9] An immunised person is a private entity who is exempt, sometimes even excluded, from the 'common law' which binds a community (Esposito 2013a, 58). Immunisation thus can be a point of personal privilege, such as an immunised political figure, or a point of absolute privation. Guest worker programmes place temporary migrant workers in the latter position. In the following section, I examine how these two sides of immunisation are operationalised by Canada's TFWP, while infusing both with a thoroughly economic theory with an economic dimension.

BIOLOGICAL LABOUR

There is a direct relationship between the gendered, heteronormative and racialised stereotypes that are rooted in biological determinism and the nature of the work performed in Canada's TFWP. Since most of the work conducted represents an extension of the traditional gendered division of labour, men and women from 'developing' nations are stereotyped as more 'naturally' fit to perform undesirable pink- or blue-collar work. The two most common fields are caregiving for pink-collar work (mostly women from the Philippines) and agricultural for blue-collar workers (mostly men from Latin America and the Caribbean) (see Table 5.2).

These dimensions have been well documented in the literature on Canada's TFWP and other similar programmes, but two points come to the fore when taking a biopolitical perspective. First, there is a direct relationship between the biologically deterministic stereotypes, the biological nature of the work and the claims to citizenship. Because temporary migrant workers are stereotyped as being more 'naturally' inclined to perform work that citizens are disinclined to perform, they are rendered biologically unsuitable for citizenship. Temporary migrant workers are treated as biological labourers who perform productive and reproductive labour that is repetitive, never ending, nonenduring yet biologically necessary for the maintenance and reproduction of the life of a nation (i.e., Hannah Arendt's notion of 'labour', 1998).[10] Vital citizens, on the other hand, conduct productive work that is postbiological, world-changing and enduring. This implies that citizens should be employed in high-skilled occupations that improve the livelihood of the nation, whereas migrant workers should only be employed in occupations that provide for the basic necessities of life.[11] This racialised, classed and gendered dichotomy has been clearly articulated since the 1970s when Canada started to redirect low-skilled migrant workers from the Global South into the TFWP. It has also

Table 5.2 Top Ten Occupational Groups for Temporary Foreign Workers (2012–2015)

Occupational Group	*2012*	*2013*	*2014*	*2015*	*Total*	*Rank*
General Farm Workers	17,286	22,946	25,791	35,999	102,022	1
Babysitters, Nannies and Parents' Helpers	16,193	13,290	12,704	3,045	45,232	2
Nursery and Greenhouse Workers	9,041	9,877	11,068	14,042	44,028	3
Harvesting Labourers	12,254	11,324	9,419	2,124	35,121	4
Food Counter Attendants, Kitchen Helpers and Related Occupations	16,197	12,776	2,381	616	31,970	5
Cooks	10,560	6,542	1,866	2,108	21,076	6
Truck Drivers	5,037	4,513	1,678	1,556	12,784	7
Food Service Supervisors	6,462	3,979	968	1,323	12,732	8
Other Performers	3,890	3,488	991	446	8,815	9
Light Duty Cleaners	3,839	2,700	735	N.A.	7,274	10
Total Positive Labour Market Impact Assessments (for all occupational groups)[a]	**199,730**	**163,035**	**104,242**	**90,211**	**557,218**	

Source: Temporary Foreign Worker Program: Annual Labour Market Impact Assessment Statistics 2008–2015—By Top Occupational Groups (ESDC 2016).

[a] These numbers are only based on the 90,211 'positive labour market impact assessments', which is a measure used in the application process for some of the new positions. They do not include workers already in the programme or those who did not require the assessment.

been reaffirmed in the 2014 splitting of the programme into the IMPs that aim to attract high-skilled foreign workers and fast-track them into becoming permanent citizens and the low-skilled TFWP that contain migrant workers in a permanent pool of temporary biological labourers. The former are treated as vital and worthy individual lives who are encouraged to settle and repopulate. The entire package of biopolitical rights offered by the Canadian state is extended to this group (health care, social insurance, public education, social and political rights, etc.), which should increase not only their individual vitality but also that of their spouses and children. Those who are treated as an indiscriminate mass of lower forms of life are not only treated as unworthy of being extended the biopolitical rights of vital citizens, but they are also formally and informally discouraged from regenerating on Canadian soil.

Second, the disempowerment of biological labourers is directly related to the empowerment of vital citizens. Vital citizens are granted extraordinary powers over biological labourers, socially, politically and economically, especially employers who can deny entry into the territory—'blacklisting', deportation, and, in the case of caregivers, impeding applications for permanent residency. This represents the near perfect neoliberal scenario because below the surface of this seemingly pure economic relationship rests a deeply politicised power imbalance between the enhanced political power of the vital citizen and the depoliticised status of the biological labourer. For the remainder of this section, I examine how these processes unfold in the most prevalent and documented feminised and masculinised fields of work: care work and agriculture.

From the nineteenth-century importation of middle class, unmarried British nannies, to the 1955 Caribbean Domestic Scheme for single women from Jamaica and Barbados, to the 1981 Foreign Domestic Movement, to the 1991 Live-in Caregiver Program, to the new 2014 Caregiver Program, there has been a shift in recruitment in temporary care work in Canada. Today, the programme targets working mothers from the Global South, mostly from the Philippines. At its peak in 2009, there were 39,559 live-in caregivers in Canada (ESDC 2014). These programmes reify the stereotype of women from the Global South as being more 'naturally', hence biologically, inclined to perform reproductive labour. In her important study of how these stereotypes are employed in Canada, for example, Geraldine Pratt notes that employing families tend to treat European nannies as 'professionals' whereas Filipina nannies are treated as 'servants', 'uncivilized' and 'childlike' (1997; 2004). Rhacel Salazar Parrenas argues that caregiver programmes exploit the stereotype of Filipina women as 'natural caregivers' (2004; 2005). Barbara Ehrenreich and Arlie Hochschild point out that these programmes create a new system of global dependency, in which the Global North is treated as the traditional male breadwinner and the Global South as the dependent female (2004). There is now a substantial literature documenting the many gendered dimensions of temporary migrant workers programmes and their social ramifications, especially the feminist literature addressing the global care deficit (Ehrenreich and Hochschild 2004; Isaksen, Devi and Hochschild 2008; Stasiulis and Bakan 2005).

Common-sense prejudices are structurally reinforced by the distinction between citizens and noncitizens. In Canada's former Live-in Caregiver Program, for example, participants were entitled to apply for permanent residency after working full-time hours for over twenty-four months within a four-year period if they lived within the private residence of their employers. Critics noted that the live-in requirement placed caregivers in a highly vulnerable position (Pratt 1997; Stasiulis and Bakan 2005; Brickner and Straehle 2010). Not only were caregivers required to live in the private home of their employer, but their employers were effectively granted the power to act as border guards who could deny their employees passage into the permanent resident stream. Such a twofold economic and political power granted employers extraordinary control over their employees, a degree of control not present in an employee-employer relationship between two citizens.[12]

On the masculinised end of the scale, we find a similar pattern of mutually reinforcing prejudices and discrimination. Canada also has a long history of hiring migrant workers to perform agricultural work. Like the caregiver programme, the agricultural programmes have been extensively studied (Basok 2002; Hennebry and Preibisch 2012; Preibisch 2010; Perry 2012). Migrant labourers sign contracts that generally range from three to eight months to work in the fields, packaging plants, and greenhouses. Today a majority of the agricultural workforce is composed of migrant workers, and approximately 96 percent are married men with families in the sending-countries. In 2012, Canada approved roughly 40,000 temporary migrant worker permits in agriculture (CIC 2012). Approximately three-quarters of temporary agricultural workers are drawn from the Seasonal Agricultural Worker Program (herein SAWP), which was established in 1966 to offset the labour shortage in the agricultural industry. The programme was initially designed to contract Jamaican labourers, but in 1974 it was extended to include Mexico. Today, it represents a patchwork of bilateral agreements between Canada and Mexico and many Commonwealth countries in the Caribbean. In the early 2000s the government also started to recruit agricultural workers in the low-skilled streams of the TFWP. In 2012, roughly one-quarter of the low-skilled agricultural workers approved to work in Canada came through this programme, with the largest numbers coming from Guatemala, Thailand, and the Philippines (McLaughlin, Hennebry and Haines 2014).

Unlike the caregiver programme, agricultural workers are limited to circular migration. Of the many contentious issues in this programme, the most controversial is how an employment contract is tied to an employer instead of the field of work. If a worker wants to change employers, he must receive permission from his employer first. Moreover, employers are able to 'blacklist' workers from the programme and to deport them (Basok, Bélanger and Rivas 2014). As with the caregiver programme, this extraordinary power imbalance places workers in highly vulnerable positions and renders them less resistant to their exploitive dirty, dangerous, and demeaning working conditions. Critics have rightly pointed out that the lack of job mobility, temporary status and the general conditions of their work contract create a working relationship that is tantamount to indentured slavery (Sharma 2001; Walia 2010)—one that is not too distant from former arrangements where feminised slaves were employed in private homes and masculinised slaves were sent to work in the fields. Only today, the master/slave dichotomy has been reconstructed in the new vital citizen/biological labourer distinction.

DEVITALISED LABOURERS

Canada's TFWP enacts a series of formal and informal immunisation mechanisms to protect the nation and the state. In the following section, I limit my analysis to a few key areas where these mechanisms have been enacted. First, I examine how the state immunises the nation from potential lethal contagion by establishing a system of surveillance to biomedically screen and practically sterilise migrant workers. Second, I outline how the state immunises itself from bearing responsibility for the vitality of temporary migrant workers by deporting unhealthy workers and, in general, exporting their reproductive costs to sending states.

The most effective way to protect the nation from potentially lethal contamination from low-skilled temporary migrant workers is to cut off their reproductive capacities. To protect the racialised *jus sanguinis* of the Canadian nation, temporary migrant workers are denied *jus domicilii* claims and their contracts are also designed to prevent potential *jus soli* claims from being made.[13] This begins with the recruitment and selection process. The Canadian government actively targets and profiles temporary migrant workers who are heterosexual, married, and have children (Hennebry and Preibisch 2012; Preibisch and Grez 2010). Unlike high-skilled workers who are given a number of incentives to permanently settle in Canada with their families, low-skilled migrant workers are effectively banned from bringing spouses or children with them because they must bear the costs of travel and settlement for their family members (Fudge and MacPhail 2009, 22). They are also required to pay international student fees for their children's education (ibid., 23). Furthermore, even spouses who manage to travel to Canada with low-skilled migrant workers are not entitled to legally work. 'These policies', note Jenna Hennebry and Kerry Preibisch, 'are principally designed to deter visa overstay or permanent settlement' (2012, 25). These and many other informal forms of discrimination faced by migrant workers have the cumulative effect of ensuring that they will establish weak roots in Canada and remain strongly attached to their home state. They are expected to work in order to enhance their lives back home, not in Canada.

This programme represents a slightly more sophisticated version of one of Canada's most historically racist immigration measures known as the 'Chinese Head Tax'. In the 1880s around fifteen thousand male labourers from rural China were employed to help build the Canada Pacific Railways (Li 1998; Bird and Short 2017, 314–16). This occurred during the height of the 'Yellow Peril' discourse, so in response to local outcries, the government passed the Chinese Immigration Act of 1885, which established a 'head tax' of $50 that was raised to $500 in 1905. The act prevented Chinese labourers from travelling with their family members. Combined with the structural and common-sense racism faced by these labourers, the act created a veritable bachelor society that was effectively sterilised, not directly but structurally and practically. Esposito calls this a process of 'sterilization at a distance' (2008, 132). This act was terminated in 1947 when people began to recognise its resemblance to Nazism.

The ideal migrant worker is a person who practices abstinence while on Canadian soil. Activists and scholars have compiled an extensive list of practices that have been employed to curb—if not outright ban—migrant workers from having sexual intercourse that could be reproductive. In addition to the effective ban on family members, employers use segregation strategies to regulate the sexual practices of temporary migrant workers. 'Historically', Preibisch and Grez argue, 'the importation of racialized male labor was also aimed at maintaining images of workers as temporary, asexual, and alien' (2010, 301). Various means have been enlisted to sterilise this population. In most workplaces, employers restrain the mobility of their workers, recruiting married workers with families, and hire workers of a single gender in a field. However, there are a few particularly conspicuous practices used in workplaces that employ men and women. First, employers profile workers on the basis of sex and nationality. Some farmers, for example, 'mix English-speaking men from one country with Spanish-speaking women from another country', not only to enhance the stereotyped images of masculine and feminised qualities of these workers but also to exploit potential prejudices held by the workers (Preibisch 2010, 418). A worker interviewed by Preibisch, for example, candidly points out that farmers tend to employ Mexican women on farms where most of the male employees are Jamaican because 'if they bring in Mexican women and Mexican men there will be trouble to pay after hours' (2010, 418). Employers even resort to implementing 'curfews, prohibiting visitors of the opposite sex, or obliging workers to inform their employers of their whereabouts when outside the farm' (ibid., 415). Furthermore, pregnancy is an '(unofficial) grounds for repatriation' and many workers are given predeparture pregnancy tests (Preibisch and Grez 2010, 305). At one point in time, 'Mexico's predeparture orientation for women involved talks that warned them not to become involved with men and even required them to sign waivers signalling their agreement' (ibid., 308). Many workers internalise these sterilisation regulations and even report other workers who breach them (ibid., 309).

In addition to the formal and informal sterilisation mechanisms, the nation is also protected by biomedical profiling procedures. Employers screen potential recruits for many different health conditions, like HIV and pregnancy, and they are subjected to regular medical examinations (Hennebry and Preibisch 2012, 25). Because nations in the Global South are perceived to have higher rates of disease due to relatively weaker health-care systems, migrant workers from the Global South are treated as a public health threat. In a separate article, the same authors point out that 'migrant workers are often healthier upon entry than their counterparts in receiving countries' due to the 'medical prescreening, selection bias and healthy behaviours, referred to as the "healthy migrant effect"' (Preibisch and Hennebry 2011, 2).

The entire package of (im)migration programmes (IMPs, TFWP, and broader immigration acts) work together as a sort of eugenics apparatus that streams potential (im)migrants into two pools: worthy vital life and unworthy biological life. The government stops short of literally sterilising temporary migrant workers, yet the effects are nevertheless the same. It is here that we should be reminded, 'the law of sterilization was in fact the first legislative measure adopted by the Nazis when in power' (Esposito 2008, 144). In a sense, then, Canada's TFWP represents a eugenics mechanism that is used to manage and tightly control the 'regeneration' of the Canadian population.

The low-skilled TFWP also has a number of built-in procedures that immunise the state from bearing responsibility for anything other than the bare existence of migrant workers. The key mechanism is the employment contract between the migrant workers and their employers. These contracts effectively immunise the state from being obliged to care for the vitality of migrant workers. The state is merely obliged to protect the interest of the employers, which means that the biological lives of migrant workers must be sustained to ensure that they can fulfil their contractual obligations, but anything above basic care remains the responsibility of their home state. Because the dearth of social welfare provisions extended to temporary migrant workers in Canada is well documented, I limit my discussion to two brief examples: poor access to health care and deportation.[14]

The employment contract requires employers to 'provide health care insurance of equal coverage to the provincial health insurance plan at no cost to the employee until the employee is eligible for coverage under the provincial health insurance plan' (Faraday 2012, 32–33). Employers must also register migrant workers in the provincial workplace insurance plan, but they cannot deduct these costs from an employee's wages. Despite this requirement, low-skilled migrant workers receive health care well below the standards expected by Canadian citizens. They are less likely to assert their biopolitical right to health care as a vital citizen would. In fact, temporary migrant labourers are placed in such precarious and exploitive conditions that their health risks are much higher than citizens (Preibisch and Hennebry 2011). Temporary foreign workers in the SAWP, for example, will use 'the public health care system for their basic needs, but many wait until they return to their home countries because they are concerned that if their employers learn they are sick or injured, they will be repatriated early and/or not be rehired in subsequent years' (Hennebry and Preibisch 2012, 26).[15]

Furthermore, since temporary migrant workers are imported to Canada for the purposes of working, they are only cared for to the extent that they can continue to work. The state is contractually immunised from bearing responsibility for their lives beyond the parameters of the contract itself. If anything were to occur that would prevent temporary migrant workers from upholding their end of the contract, then the contract would be terminated and the state would no longer be politically and legally obliged to provide further services. Should an accident occur that would render migrant workers incapable of working on a long-term basis, employers immediately deport them (Basok 2002; Hennebry and Preibisch 2012). This represents an uneven trade-off between sending and receiving states. Importing states exploit and appropriate their labour power, but they export the costs that are required to maintain the 'heath care for injured and ill returning migrant workers' (Perry 2012, 192). Exporting states bear the costs of caring for the vital elements of their citizens, ranging from their education, health care, social assistance, and old-age security (Sassen 1988). This arrangement represents a near perfection of neoliberalism. As mere economic entities with little political, social and civil rights, low-skilled migrant workers are minimally cared for to the extent that they are able to uphold the terms of the contract. As nonvital, temporary workers, their lives are only worth protecting to the extent to which they continue to contribute to the economy. Their lives, in other words, are not important in themselves, but for their use-value. This arrangement, in short, contractually immunises the state from bearing responsibility for anything other than the bare life of temporary migrant labours. The state will help them exist but not flourish. In other words, this apparatus is being used for bioeconomic ends just as much as it is for biopolitical ends. In the end, it establishes the ultimate neoliberal system of workfare, not welfare.[16]

CONCLUSION

There is no question that the cumulative effects of the biopolitical discrimination examined in this chapter would never be tolerated by the vital citizens of Canada, especially not by middle- and upper-class white settler citizens. Unfortunately, Nandita Sharma's statement from more than a decade and a half ago continues to ring true today:

> imagine the outcry if all Canadian citizens who were accountants, university professors or autoworkers were legally indentured to their employers and faced harsh penalties for quitting or changing their jobs. Imagine, too, the outcry if all were denied access to minimum employment standards, ability to organize into unions or social programs and services. (2001, 427)

Imagine the outcry, we might add, from vital citizens if they were denied their biopolitical rights as biological labourers are in the TFWP.

There are two points I want to end on. First, critical scholarship in biopolitics must seriously address the intertwinement of the political and the economic in biopolitics. This perspective carries us beyond the various moves to either bracket off the economic or to examine the parallelism between the economic and the political, as if they are separate and autonomous spheres. Too much scholarship remains trapped in examining the division between proper and improper life on a normative level. This chapter employed the case study of Canada's TFWP to demonstrate how the normative division is embedded in a broader, global biopolitical economy of life. The biopolitical does not merely discriminate between proper and improper lives that are then taken advantage of and exploited in a second economic dimension. Discrimination is reinforced by exploitation, and exploitation is reinforced by discrimination—they are mutually reinforcing, or coproductive, dimensions of a broader series of thoroughly integrated processes. It is the combination of the bioeconomic and the biopolitical that we need to start focusing on. Life itself is appropriated, divided up and asymmetrically distributed in our now-global biopolitical economy. Life is not being used as a metaphor in these sentences but as a good that is slowly being commodified, converted into an exchange value with its own currency and redistributed on a global scale. We are aware that other forms of life, from flora to fauna, from 'natural' life to biologically engineered life, have been commodified and distributed in the bioeconomy of contemporary capitalism, but little has been written that account for the distribution (not just division) of the lives of so-called humans. Much has been written about the biological (survival, defence, protection, reproduction, eugenics, etc.), but not enough about the vital. By focusing on the distribution of vitality, we can start to focus on a much larger problem.

Second, temporary migrant workers are not private, economic entities but active contributors to the common. It is the duty of those who have more access to the common to recognise their contribution and thus to struggle for a redistribution of common goods and services in a much more equitable manner. The vitality of all the lives contributing to the common must be affirmed; otherwise we will remain trapped in a negative biopolitical system of differential inclusion, which has negative consequences for everyone. If we are to rethink the common on a global scale, it must begin with an affirmative notion of biopolitics. This common must be open, inclusive and without borders. The common is the space of affirmation, rather than negation, of everyone's vitality. Many organising efforts today are actually unified around the demand that temporary migrant workers should be recognised and treated as vital lives. Migrant worker activists, organisations representing them, trade unions and researchers have been fighting to counter this negative model. From legal battles to provide access to employment insurance, to establishing service centres in regions with a high concentration of temporary migrant workers, to unionisation efforts, to providing adequate access to health care, to street protests against deportation, to calls for public inquiries regarding workplace deaths and injuries, what is starting to emerge is a collective effort to establish a model of biopolitics that affirms temporary migrant workers as vital singular and plural lives.

NOTES

I would like to thank Hannah Richter for inviting me to participate in the original workshop at the University of Kent that she organised in December 2015, for her thorough feedback on this chapter, and for all the time she has dedicated to this collected book. Thank you also to Maria Fannin for your thoughtful review of this chapter and to Kathy Bischoping, James Chamberlain and Jon Short who reviewed earlier versions of this chapter.

The epigraph is from a statement by Conservative Member of the Canadian Parliament John Williamson on labour shortages in meatpacking and processing plants in New Brunswick, Canada. The implication is that temporary foreign workers (i.e., 'foreign' and 'brown' people) are forcing 'white' Canadians out of paid employment and into government welfare programmes.

1. Timothy Campbell was the first to fully examine the ways that contemporary biopolitical philosophers have incorporated the Heideggerian theme of the proper into

their analyses in *Improper Life* (2011). In *Containing Community*, I examined various political, economic, ethical and normative dimensions of the 'dispositif of the proper' (Bird 2016).

2. For the purposes of clarity, I employ the term 'migrant worker' in this chapter. It should be noted though that this is a highly contentious term, given that it connotes a sense of mobility and freedom, which—as I show in this chapter—is not the case for most migrant workers working in Canada in these programmes.

3. I used the same demographic data in a paper I cowrote with Jon Short called 'Cultural and Biological Immunization: A Biopolitical Analysis of Immigration Apparatuses' (Bird and Short 2017). We conducted a genealogical analysis of what we call the two stages of immigration policies in Canada: 'crude immunization stage' and 'sophisticated immunization stage'. In this chapter on Canada's guest worker programmes, I conduct a more detailed analysis of the practical implications and practices of the second sophisticated immunization stage.

4. For an extensive study of the child deportation or 'Home Children' scheme, see Parker 2010.

5. There is now an extensive literature on this topic in Canada, some of which I cite in this paper. It is important to note that these contradictory policies were first fully elaborated by Nandita Sharma in the early 2000s (Sharma 2001; 2006).

6. Canada's TFWP is actually a patchwork of many different programmes and streams, some of which are products of bilateral and multilateral trade agreements. Accounting for the patchwork of programmes, distinctions and the sheer demographics for migrant workers in Canada has been a difficult task for scholars and activists working in this field—a complexity many critics argue that protects the government from public scrutiny. It is beyond the scope of this chapter to provide a clear diagram of these programmes. My intention is merely to examine them from a global perspective.

7. These calculations are based on the figures from 2005, the last year in which the Liberal Party of Canada was the governing party. During the Jean Chrétien/Paul Martin Liberals' reign from 1993 to 2005, the number of temporary migrant workers grew by a total of 130.4 percent, while the number of permanent residents merely increased by a total of 102.9 percent.

8. To break this down, 11,929 (Level 0—Managerial), 32,561 (Level A—Professional), 20,760 (Level B—Skilled and Technical), 9,292 (Level C—Intermediate and Clerical), and 3,397 (Level D—Elemental and Labourers) (CIC 2012).

9. Community and immunity derive from the Latin term *munus* (*communitas* and *immunitas*). *Munus* connotes a gift, duty and obligation (Esposito 2010, 4; 2013a, 58). In *communitas* members are exposed to each other, such that each is bound to give back to the community and care for one another. Immunized communities represent the opposite. They become 'small homelands that break apart the universal idea of community into many small, unified micro-communities, opposed by definition to each other by ethnic, linguistic, and cultural identity' (2010, 54–55). For a detailed account of Esposito's notion of the *munus*, see Esposito 2010, especially the introduction and appendix, Bird 2013, Richter 2016; Weir 2013.

10. This argument is similar to the one advanced by Dauvergne and Marsden who draw from Hannah Arendt to demonstrate how temporary migrant labour programmes 'reduce people to labour alone' (2014, 235). In this model, 'one's *existence* as citizen is contingent upon one's value as labourer' (ibid., 239). This model, they argue, 'hollows the idea of citizenship and deprives it of meaning' (ibid.).

11. This is a slight variation on Rose's term 'biological citizen'. In 'Biological Citizenship', for example, Rose and Novas claim to be using the term 'descriptively, to encompass all those citizenship projects that have linked their conceptions of citizens to beliefs about the biological existence of human beings, as individuals, as families and lineages, as communities, as populations and races, and as a species' (2007, 440). I use the term 'biological labourer' to reflect a biopolitically informed reading of the literature on temporary migrant workers in Canada, specifically the feminist and antiracist literature.

12. In 2014, the government split the programme into two streams: Caring for Children Pathway and Caring for People with High Medical Needs Pathway. The 'live-in' requirement has been dropped and there are now caps on the number of applications per year at 2,750 per stream (ESCD 2014). Nevertheless, caregivers must still work for a minimum of two years before they are eligible to apply for permanent residency, so the presence of the border distinguishing between vital citizens and biological labourers remains.

13. Under Canada's point system, most low-skilled temporary migrant workers would not qualify for permanent residency and by signing a contract they are also prevented from applying for it. There are some exceptions to this rule, such as the Caregiver Program (3,690 workers transitioned from temporary to permanent residency in 2012) and the Provincial/Territorial Nominee Program that allows governments to bypass the points system if they can demonstrate economic and labour needs (CIC 2012).

14. Preibisch and Hennebry list several factors that make low-skilled migrant workers more vulnerable: frequent migration; status tied to employers; dangerous working conditions; insufficient health screening; barriers to health care and insurance; lack of independent monitoring; insufficient safety equipment and training; little information, representation and support; poor and underregulated housing; social exclusion and isolation; and linguistic and cultural differences (2011, 3).

15. Based on a survey of over six hundred temporary workers in the SAWP, Hennebry found that '[w]hen temporary migrants experience symptoms or injuries, they face significant barriers to accessing adequate care, including dependence on the employer for transportation and access to health insurance, language barriers, fear of reprimand, termination of their contract and repatriation' (2010, 75). Less than one-quarter of workers who experienced health problems due to work, she found, had actually seen a doctor in Canada.

16. Chamberlain (2014) uses Esposito's theory to examine how this distinction is employed in the case of undocumented workers in the United States.

BIBLIOGRAPHY

Arendt, Hannah. 1998. *The Human Condition*. Chicago: University of Chicago Press.

Basok, Tanya. 2002. *Tortillas and Tomatoes: Transmigrant Mexican Harvesters in Canada*. Montreal: McGill-Queen's University Press.

Basok, Tanya, Danièle Bélanger and Eloy Rivas. 2014. "Reproducing Deportability: Migrant Agricultural Workers in South-Western Ontario." *Journal of Ethnic and Migration Studies* 40 (9): 1394–1413.

Bigo, Didier. 2002. "Security and Immigration: Toward a Critique of the Governmentality of Unease." *Alternatives* 27: 63–92.

Bird, Greg. 2013. "Roberto Esposito's Deontological Communal Contract." *Angelaki* 18 (3): 33–48.

Bird, Greg. 2016. *Containing Community: From Political Economy to Ontology in Agamben, Esposito, and Nancy*. Albany: State University Press of New York.

Bird, Greg, and Jon Short. 2017. "Cultural & Biological Immunization: A Biopolitical Analysis of Immigration Apparatuses." *Configurations* 25 (3): 301–26.

Brickner, Rachel K., and Christine Straehle. 2010. "The Missing Link: Gender, Immigration Policy and the Live-in Caregiver Program in Canada." *Policy and Society* 29 (4): 309–20.

Campbell, Timothy C. 2011. *Improper Life: Technology and Biopolitics from Heidegger to Agamben*. Minneapolis: University of Minnesota Press.

Chamberlain, James A. 2014. *The Power of Work: Rethinking the Politics of Immigration*. (Working Paper).

CIC (Citizenship and Immigration Canada). 2011. "Facts and Figures 2011, Immigration Overview: Permanent and Temporary Residents." http://publications.gc.ca/collections/collection_2013/cic/Ci1-8-2011-eng.pdf. Accessed August 2014.

CIC (Citizenship and Immigration Canada). 2012. "Facts and Figures 2012, Immigration Overview: Permanent and Temporary Residents." http://publications.gc.ca/collections/collection_2013/cic/Ci1-8-2012-eng.pdf. Accessed 14 January 2015.

CIC (Citizenship and Immigration Canada). 2014. "Facts and Figures 2014, Immigration Overview: Permanent and Temporary Residents." http://publications.gc.ca/collections/collection_2015/cic/Ci1-8-9-2014-eng.pdf. Accessed 1 April 2016.

Dauvergne, Catherine, and Sarah Marsden. 2014. "The Ideology of Temporary Labour Migration in the Post-Global Era." *Citizenship Studies* 18 (2): 224–42.

Ehrenreich, Barbara, and Arlie Russell Hochschild, eds. 2004. *Global Woman: Nannies, Maids, and Sex Workers in the New Economy*. New York: Henry Holt.

ESDC (Employment and Social Development Canada). 2014. "Improving Clarity, Transparency and Accountability of the Temporary Foreign Worker Program." https://www.canada.ca/en/employment-social-development/services/foreign-workers/reports/2014/lmia-annual-statistics.html. Accessed 14 January 2015.

ESDC (Employment and Social Development Canada). 2016. "Temporary Foreign Worker Program: Annual Labour Market Impact Assessment Statistics 2008–2015—By Top Occupational Groups." http://www.hrsdc.gc.ca/eng/jobs/foreign_workers/lmo_statistics/annual-top-occs.shtml. Accessed 12 February 2017.

Esposito, Roberto. 2008. *Bíos: Biopolitics and Philosophy*, translated by Timothy Campbell. Minneapolis: University of Minnesota Press.

Esposito, Roberto. 2010. *Communitas: The Origin and Destiny of Community*, translated by Timothy Campbell. Stanford: Stanford University Press.

Esposito, Roberto. 2011. *Immunitas: The Protection and Negation of Life*, translated by Zakiya Hanafi. Cambridge: Polity.

Esposito, Roberto. 2013. *Terms of the Political: Community, Immunity, Biopolitics*, translated by Rhiannon Noel Welch. New York: Fordham.

Faraday, Fay. 2012. "Made in Canada: How the Law Constructs Migrant Workers' Insecurity." Metcalf Foundation. http://metcalffoundation.com/publications-resources/view/made-in-canada. Accessed August 2014.

Foucault, Michel. 1981. *The History of Sexuality I*, translated by Robert Hurley. London: Penguin.

Foucault, Michel. 2003. *Society Must Be Defended: Lectures at the Collège de France 1975–1976*, translated by David Macey. New York: Picador.

Fudge, Judy, and Fiona MacPhail. 2009. "The Temporary Foreign Worker Program in Canada: Low Skilled Workers as an Extreme Form of Flexible Labour." *Comparative Labor Law and Policy Journal* 31 (5): 5–45.

Hennebry, Jenna L. 2010. "Not Just a Few Bad Apples: Vulnerability, Health and Temporary Migration in Canada." *Canada Issues/Thèmes Canadiens*, Spring: 73–78.

Hennebry, Jenna L., and Kerry Preibisch. 2012. "A Model for Managed Migration? Re-Examining Best Practices in Canada's Seasonal Agricultural Worker Program." *International Migration* 50 (s1): 19–40.

Isaksen, Lise Widding, Sambasivan Uma Devi and Arlie Russell Hochschild. 2008. "Global Care Crisis: A Problem of Capital, Care Chain, or Commons?" *American Behavioral Scientist* 52 (3): 405–25.

Levinson, Brett. 2010. "Biopolitics in Balance: Esposito's Response to Foucault." *New Centennial Review* 10 (2): 239–62.

Li, Peter S. 1998. *The Chinese in Canada*, 2nd ed. Toronto: Oxford University Press.

Mazer, Katie, and Josie Baker. 2015, March 27. "Migrants in the Maritimes: How to survive in a remittance economy." http://rabble.ca/news/2015/03/migrants-maritimes-how-to-survive-remittance-economy. Accessed 26 March 2018.

McLaughlin, Janet, Jenna Hennebry, and Ted Haines. 2014. "Paper versus Practice: Occupational Health and Safety Protections and Realities for Temporary Foreign Agricultural Workers in Ontario." *Perspectives Interdisciplinaires Sur Travail et La Santé* 16 (2): 1–12.

Medovoi, Leerom. 2007. "Global Society Must Be Defended: Biopolitics without Boundaries." *Social Text* 25 (2): 53–79.

Parker, Roy. 2010. *Uprooted: The Shipment of Poor Children to Canada, 1867–1917*. Bristol: Policy Press.

Parrenas, Rhacel S. 2004. "The Care Crisis in the Philippines: Children and Transnational Families in the New Global Economy." In *Global Woman: Nannies, Maids and Sex Workers in the New Economy*, edited by Barbara Ehrenreich and Arlie Russell Hochschild, 39–54. New York: Holt Paperbacks.

Parrenas, Rhacel S. 2005. *Children of Migration: Transnational Families and Gender Woes*. Palo Alto, CA: Stanford University Press.

Perry, J. Adam. 2012. "Barely Legal: Racism and Migrant Farm Labour in the Context of Canadian Multiculturalism." *Citizenship Studies* 16 (2): 189–201.

Pratt, Geraldine. 1997. "Stereotypes and Ambivalence: The Construction of Domestic Workers in Vancouver, British Columbia." *Gender, Place, and Culture* 4 (2): 159–78.

Pratt, Geraldine. 2004. *Working Feminism*. Philadelphia: Temple University Press.

Preibisch, Kerry. 2010. "Pick-Your-Own Labor: Migrant Workers and Flexibility in Canadian Agriculture." *International Migration Review* 44 (2): 404–41.

Preibisch, Kerry, and Evelyn Encalada Grez. 2010. "The Other Side of El Otro Lado: Mexican Migrant Women and Labor Flexibility in Canadian Agriculture." *Signs* 35 (2): 289–316.

Preibisch, Kerry, and Jenna Hennebry. 2011. "Temporary Migration, Chronic Effects: The Health of International Migrant Workers in Canada." *Canadian Medical Association Journal*: 1–6.

Richter, Hannah. 2016. "Beyond the 'Other' as Constitutive Outside: The Politics of Immunity in Roberto Esposito and Niklas Luhmann." *European Journal of Political Theory*: 1–20 (online first) http://journals.sagepub.com/doi/full/10.1177/1474885116658391. Accessed 26 March 2018.

Rose, Nikolas. 2007. *The Politics of Life Itself: Biomedicine, Power, and Subjectivity in the Twenty-First Century*. Princeton, NJ: Princeton University Press.

Rose, Nikolas, and Carlos Novas. 2007. "Biological Citizenship." In *Global Assemblages: Technology, Politics and Ethics as Anthropological Problems*, edited by Aihwa Ong and Stephen Collier, 439–63. Oxford: Blackwell.

Sassen, Saskia. 1988. *The Mobility of Labor and Capital: A Study in International Investment and Labor Flow*. New York: Cambridge University Press.

Sharma, Nandita. 2001. "On Being Not Canadian: The Social Organization of 'Migrant Workers' in Canada." *Canadian Review of Sociology/Revue Canadienne de Sociologie* 38 (4): 416–39.

Sharma, Nandita. 2006. *Home Economics: Nationalism and the Making of "Migrant Workers" in Canada*. Toronto: University of Toronto Press.

Stasiulis, Daiva Kristina, and Abigail Bess Bakan. 2005. *Negotiating Citizenship: Migrant Women in Canada and the Global System*. Toronto: University of Toronto Press.

Walia, Harsha. 2010. "Transient Servitude: Migrant Labour in Canada and the Apartheid of Citizenship." *Race & Class* 52 (1): 71–84.

Weir, Lorna. 2013. "Roberto Esposito's Political Philosophy of the Gift." *Angelaki* 18 (3): 155–67.

Chapter 6

The Biopolitics of Donation

Gender, Labour and Motherhood in the Tissue Economy

Maria Fannin

The proliferation of neologisms that take Michel Foucault's articulation of biopolitics as their starting point—biocapital, biolabour, biovalue, bio-economy and so on—are evidence of the profoundly influential place of this concept for analysing contemporary political cultures and for writing a 'history of the present'. In this chapter I consider the role of the life sciences as biopolitical sites and focus on how new spaces and subjects of the 'bio-economy' extend Foucault's articulation of biopolitics into new domains. The contemporary biopolitics of the life sciences encompass the proliferation of reproductive technologies, exchanges of bodily materials, policies to control fertility at the level of the population *and* efforts to encourage the cultivation of individual responsibility for one's reproductive capacities. I focus on how the donor and the clinical labourer feature in feminist accounts of the contemporary life sciences to consider how attention to gender invites a rethinking of the biopolitics of reproduction in this field. The aim is thus to highlight how attention to political life requires attention to fertility, pregnancy and birth, as well as investment of capital, labour and value in these reproductive dimensions of the living being.

Indeed, contemporary developments in the life sciences, as this chapter argues, are co-constitutive with new forms of discipline as well as the production of new subjects and spaces of governmental control. These include efforts to control and survey the molecular and population movements of disease and the generation of new communities of medical subjects. There is now an extensive literature on the intersections between what Foucault described as the 'politics of life' and the late twentieth- and early twenty-first-century dynamics of what has been termed the 'bioeconomy' (Rajan 2006; Larsen 2007; Rose 2007; Helmreich 2008; Goven and Pavone 2014). Broadly speaking, the bioeconomy has been characterised as the sector of the global economy that specifically takes life as its object, and has been variously identified in literatures on political economy and public policy with the emergence of the genetic manipulation of plant and animal breeding, the expansion of the pharmaceutical industry, the growth in gene-based biotechnologies and the role of the life sciences as engine for regional and national economic development (Birch 2012; Birch and Tyfield 2012). The hype surrounding biotechnology in particular in the late twentieth century spurred a range of studies assessing the bioeconomy's capacities for growth, its institutional dynamics and legal infrastructures, and the blurring of public and private finance that characterised research and development in the sector. More closely following Foucault's rendering of biopolitics as the convergence of the politics of life alongside forms of economic rationality, Kaushik S. Rajan (2006) identifies the life sciences as a key site for an epistemological transformation in the relationship between knowledge, bodies and power as well as a new phase of neoliberal capitalist political economy he terms 'biocapital'. Foucault's own work on the biopolitics of population does not explicitly anticipate the new technological formations surrounding the life sciences nor envisage their evolution over time. However, Foucault's articulation of biopolitics and economy suggest that the bioeconomy can be fruitfully conceptualised, not as a new technology-led phase of biopolitics but rather as a set of 'governmental technologies structuring the political government of life' (Larsen 2007, 21). In this way, biopolitics and the biotechnological economy are co-constituted and manifest the longer conjuncture of 'life' and 'economy' Foucault describes in his writings on biopower and the 'truth-telling' functions of economic rationality.

Foucault's theorisation of power, biopolitics, sexuality, self-cultivation and madness continue to inspire feminist engagement with his work (Allen 2013). As feminist and postcolonial scholars have noted, Foucault's own articulation of the racialised and gendered dimensions of biopolitics was relatively scant (McNay 2009; Stoler 1995). Recently published work carried out with Arlette Farge on letters selected from French pre-Revolutionary archives points to Farge and Foucault's shared sensitivity to how family dynamics in this period were shaped by expectations regarding the differences and relationships between men and women. Analysis of this archive, consisting of letters written to the king by poor and working people requesting intervention into family conflicts between parents and children and between wives and husbands, reveals dynamics that anticipate the disciplinary mechanisms described in Foucault's account of nineteenth-century prisons. But it also provides an extraordinary insight into a period prior to the establishment of the Napoleonic Civil Code and its instantiation of the authority of the husband or father as head of household:

> the place of reproduction would from then on be managed by the male world . . . a domestic space where men would naturally lay down the rules began gradually to take shape. Detached from the field of public events, married life would obligate women to exit the stage. Between State and women, in this specific case, there was reciprocity no longer; their spaces had been severed in a quasi-definitive way. It was men who would assure the link between the two and who would, for this reason, relegate women to the confined space of private life. (Foucault and Farge 2016, 49)

This account of the management of reproduction within the private life of the family and the severing of a direct link between 'State and women' would be reflected in Foucault's articulation of the role that the 'hystericisation of women's bodies' played in managing and administering women's sexuality through medicine, hospitals and the 'psy' disciplines. It would also prefigure a new role for the biopolitical state and economy in defining fertility, pregnancy and birth as population-based concerns, and where the gradual loosening of restrictions around these processes would require new forms of responsibility on the part of individuals.

However, much of the early feminist and gender-sensitive scholarship on population control, reproductive technologies, transformations in women's health care and access to abortion or contraception only obliquely engages with Foucault's analysis of biopolitics, citing the usefulness of Foucault's concepts while noting the absence of sustained engagement with the specificities of disciplining and governing women's bodies on his part. This may be in part due to readings of Foucault's published work on 'sex' through the lens of the biopolitics of population or, in parallel but rarely overlapping literatures, as a question of identity and the making of the self. Instead, what is needed is greater attention to what Penelope Deutscher (2012) calls the 'procreative hinge' at work in Foucault's writings on biopolitics in order to enable analysis of the specificities of abortion laws, population policy, alongside women's movements for 'rights to the body', eugenic feminism and the governmental practices of reproductive self-fashioning. For example, while framing their work to varying degrees as focused on biopolitics or the bioeconomy, feminist scholars examine how new medical technologies have transformed women's experiences of fertility, pregnancy and birth (Petchetsky 1987; Shildrick 1997; Weir 2006). Technologies to enable in vitro fertilisation (IVF) have multiplied the routes to pregnancy, and the emergence of contractual forms of surrogacy, particularly by would-be parents in the United States, has engendered new global dynamics of exchange of biological materials and children across bodies and borders (Parry et al. 2015). New 'tissue economies' have emerged involving materials produced during pregnancy that are typically discarded only to be recirculated into research settings as resources for stem cell research and regenerative medicine (Waldby and Mitchell 2006). All of these technologies, implicated in efforts to target the health of the population as well as in the extension of extend forms of reproductive self-fashioning, have invited a reassessment of the gendered and racialised dimensions of the contemporary biopolitics of reproduction and the life sciences.

Despite the incisive critiques of how the technologisation of reproduction both expands the possibility of parenthood and generates new forms of risk, surveillance and bodily discipline, early examination of these technologies in feminist literatures took place without any explicit conversation with broader work on biopolitics as a political concept. It is only in the past decade that these technologies have been the focus of sustained efforts to conceptualise the relationship between the biopolitics of reproduction and broader economic processes in the emerging capitalist 'bioeconomy' (see, for example, Larsen 2007). Although the biopolitical concern for promoting the life of a population is intimately bound with the reproduction of population, it would fall to later scholars to articulate clearly, for example, how the relationship of National Socialism to eugenic theory and practice targeted women's reproductive capacities as central to the making of race (Deutscher 2010). Within scholarship on the history of feminism, there is disagreement over the extent to which population-based efforts to manage and control subjects were indebted to and supported by feminist and women's movements. Drawing on Foucault's work, philosophers and historians of biopolitics have since explored the relationship between the popularity of eugenics in the early to mid-twentieth century and political movements such as feminism (Richardson 2003; Bashford 2014). Eugenic policies exerted new forms of control over women's bodies even while they were embraced by scientists, feminists and politicians as the means of liberating poor women from the burdens of childbearing.

Foucault's discussion of biopolitics as a strategy targeting the creation and identification of populations also implicates measures aimed at controlling and disciplining individual bodies and selves through modes of engendering personal responsibility for one's health (Petersen 1996; McNay 2009). This conjuncture, between efforts to manage the health of the population and the practical means by which these efforts target individual bodies, is the focus of recent feminist interpretations of Foucault's work that identify reproductive processes as central to population-based biopolitics, and attention to the regulation and control of women's reproductive capacities as the means by which biopolitics operates (Fannin 2013). As Catherine Mills notes, Foucault directly identifies the 'socialization of procreative behavior' (Foucault 1990, 105; cited in Mills 2017, 282) as a key site for the regulation of sexuality combining juridical intervention into the family form with a bodily 'economy of pleasure'. This involved the introduction of a whole set of concerns for the population and the individual that Foucault identifies as medical, social and political around the conduct of procreation, in particular the concern for birth control practices. Feminist scholarship has taken up these insights to highlight how efforts to govern the reproduction of a population have simultaneously targeted a collective as well as individual subjects and bodies, examining, for example, the coordination of efforts by eugenic feminists and public health advocates to argue for the benefits of birth control for the 'life of the body and the life of the species' (Deutscher 2012, 124; see also Deutscher 2008; 2010).

The transformation of reproductive practice and the reworking of the boundaries between social responsibility for public health; the lifting of restrictions on contraception, abortion, and reproductive technologies; and individual burdens (women's responsibility to make 'wise' reproductive choices), once only obliquely linked to the emergence of the life sciences as engine of investment, economic development, and a new 'era' of capitalism, are indeed central to it. In the following sections, I examine the biopolitics of technologies that take the generative capacities of the pregnant and birthing body as their target and object. Feminist scholars have reworked Foucault's account of biopolitics to elaborate on the silences in his thinking in relationship to the sexed duality of 'species-being', although not without disagreement over the relevance of gender as an analytic. For example, recent work by Jemima Repo (2015) suggests that the concept of gender has little place in a Foucauldian analysis of biopolitics; rather, gender itself must be the subject of genealogical critique. Repo's valuable intervention into feminist theoretical approaches to biopolitics carries out an incisive historical analysis of gender as a discursive and material assemblage of institutions and strategies for ordering bodies and sexes.

She calls for feminist theorists to reconsider gender as a concept and mode of analysis, inviting feminists to delineate carefully, with attention to the historicity of 'gender' as an analytic and political tool, how this concept is mobilised in their own work. Following Repo's insistence on making the terms of the use of 'gender' clear, this chapter draws on gender as an analytic that reveals how bodily materials are invested with different values and how forces govern the vital politics of donations from sexed reproductive bodies in specific ways, but also returns to Foucault's articulation of 'sex' in his writings on biopolitics in order to explore more thoroughly the significance of gendered dynamics of donation, and specifically donation of 'reproductive' tissues, as biopolitical sites. In the following sections, I examine these gendered dimensions of the biopolitics of donation through the figures of the pregnant and birthing donor, the performative 'clinical labourer', and the governmentalised maternal subject. Taken together, they reveal how tissue donation in the bioeconomy operates by targeting the life processes of fertility, pregnancy and birth in new and important ways.

'BETWEEN THANATOPOLITICS AND BIOPOLITICS': HUMAN TISSUE DONATION

The rhetoric surrounding human tissue, blood and organ donation suggests new configurations of what Nikolas Rose and Carlos Novas (2004) term 'biological citizenship', configurations of belonging that take the biological being of individuals, populations, races and species as central to citizenship projects. The centrality of donation to citizenship projects was the subject of Richard Titmuss's (1970) landmark text on the social policy implications of donation. Subsequent work exploring the practices of inclusion and exclusion of donors from national blood donation systems highlights how citizenship is also a biopolitical project in which some bodily 'gifts' are deemed to be essential to the life of the polity and others viewed as harmful—for example, in policies temporarily or permanently prohibiting blood donation from men who have sex with men (MSM) (Martucci 2010; Bennett 2009). This exclusion, and the claims to forms of biological citizenship arising from efforts to contest it, shape the formation of new kinds of '*vital publics* . . . defined by those who may give, receive, or otherwise benefit from, blood and other biological material, the putative substances of "life"' (Strong 2009, 172–73, emphasis added). Yet the figure of the donor in these accounts of vital or biological citizenship is often undifferentiated by gender, despite the clear disparities in many donation systems between who gives and who receives.

In the United Kingdom, more women donate blood and organs than men, yet with regard to organ transplantation, more men receive these organs than women (Steinman 2006). Donation is often presented as an act of care and the act of giving part of oneself to an unknown other, yet the reasons for this difference with respect to stereotypes of donation as nurture and care or through evidence of disparities in the prevalence of diseases necessitating transplant are not very well substantiated. The extent to which 'gender' and 'biology' operate in the acquisition and distribution of organs, for example, is not entirely understood. However, in the case of blood donation in the United Kingdom, the presumption that women donors may have been pregnant directly determines how plasma, one of the components of whole blood, is valued in the national blood system. Pregnant—or previously pregnant—bodies are deemed to pose risks to the national blood supply yet the routine disposal of plasma from women's donated blood is nowhere communicated to donors (Kent and Farrell 2014). Estimations of the riskiness of bodies and the perceived threat to the population from donated blood are shaped by gendered presumptions about sexual behaviour and the role of donation as a form of biological citizenship. Gender also figures into the collection and value attributed to donated reproductive tissues that cannot easily be ascribed to the physiological differences between male and female bodies.

In Rene Almeling's (2009; 2011) research on the 'donation' of sperm and eggs for use in California IVF clinics, the economic and moral value attributed to these materials was differentiated by gendered presumptions that sperm donors were oriented towards market-based calculations of compensation and lack of concern about the fate of their donated materials, while egg donation was imagined by donors and clinics as an act of generosity to help recipients create families. Almeling (2009, 37) notes that despite the language of 'donation', exchanges of sperm and eggs in the United States operate as a 'twenty-first century medical market in reproductive goods'.

The donation practices surrounding tissues generated by pregnancy, including foetal tissue, placenta, umbilical cord blood and breast milk, further highlight the importance of understanding donation as a practice shaped by expectations regarding the bodies of men and women and the relations between them (Kent, Fannin and Dowling 2017). Unlike the donors of embryos created from IVF procedures and redirected to research use, donors of foetal tissue derived from abortion were presumed to be uninterested in the fate of the donated tissue and less likely to be provided with details of the specific research using their donated tissue (Kent 2008; Pfeffer 2009). This suggests a further differentiated biopolitics of reproductive tissue donation, in which pregnancy is differentiated as socially and individually 'valued' or 'unwanted'. Placental tissue and cord blood donation, whether for research or for public or private storage and use, are similarly shaped by gendered presumptions of moral and economic value: because materials donated from pregnancy are routinely defined as waste, finding another use for these materials that attributes use and exchange value to them is often seen as ethically unproblematic (Fannin and Kent 2014). This obscures the broader biopolitics of using human tissues as forms of 'biocapital' in which the molecular life processes of cellular growth and development are invested with new forms of agency and moral, economic and social value (Santoro 2011; Waldby and Mitchell 2006; Brown and Kraft 2006).

The gendering of donation is also at work in the maternal-child relation, in which fluids such as breastmilk pass from one body to another but are excluded from conceptual considerations of donation. Mothers are not identified as donors to their own children, despite the fact that if they were to breastfeed another's child, they would be viewed as donors and the act of feeding as a donation. Because the child is seen as fully dependent on the maternal body (or a substitute for this body) for survival and sustenance, the movement of fluids from the maternal to the child's body is not identified as a donation. Rather, this exchange between bodies is viewed to meet the child's basic needs and although it is a valued act, it is often circumscribed within the intimate relations and space of the family, rather than viewed as a civic act of giving. This echoes Almeling's work discussed earlier in which selling eggs is viewed as an act of care, often located within the family, while selling sperm is more clearly identified as a means of entering into an exchange on a (public) market. The paradigmatic figure of the donor participating in a civic and altruistic act of solidarity through their act of donation is revealed as an undifferentiated figure, unmarked by gendered expectations or forms of bodily difference that are in fact central to the structure of donation but not acknowledged as such. The distinction between the civic donor and the intimate giving and receiving of bodily material in the 'private' realm thus calls into question the significance of the indistinction diagnosed by theorists of biopolitics like Giorgio Agamben between private life, *zoe,* and the public figuring of life as *bios* within contemporary biopolitics, while also underscoring the limitations of Agamben's inattention to the gendered dimensions of this distinction (Deutscher 2008).

Penelope Deutscher writes that 'one can think about women's reproductive life (women's lives as valued, problematised and rendered significant in their capacity to bear and produce new life, and as exposed to injunction or loss of life for that same capacity) as occupying a significant position at the nexus of biopolitics and thanatopolitics' (2010, 221). Deutscher's invoking of the powers to make live *and* make die raises the question of how population-based policies seeking to enhance and promote the lives of some were historically accompanied by efforts to limit and negate the lives of others. Deutscher turns to Roberto Esposito's immunitary theory of biopolitics to argue that 'reproductive politics were concurrently thanatopolitical', which is to say that the politics of protecting and enhancing women's reproductive lives took place on the very same terrain—of debates over state intervention and reproductive autonomy—as that which sought to end women's reproductive lives, through the risk of death from unsafe abortion, the establishment of death penalties for procuring abortion or forced sterilisation (ibid., 226).

Deutscher's conceptual framing of women's reproductive lives as invested with both thanatopolitics and biopolitics suggests that the social exhortation to view one's body and body parts as potentially valuable for donation also has a thanatopolitical counterpart. The thanatopolitics of donation, envisioned biopolitically as a means of generating solidarity or exchange, operates in the political cultures that attempt to negate life through state-sanctioned violence, environmental devastation and the investment of some lives with more moral value than others. The figure of the human tissue donor is shaped by exhortations to give 'the gift of life' and to make 'wise use' of bodily materials while living bodies are increasingly subject to the threat of violence and death. This suggests that one of the paradigmatic biopolitical subjects of the twenty-first century may be the donor who is solicited to give in the name of saving life but whose life as a living being is abandoned and threatened by planetary and geopolitical crises and the abrogation of neoliberal states of responsibilities for collective health, leaving individuals with the responsibility of caring for their own lives.

Feminist scholarship on the donation economy draws attention to how tissues derived from pregnant bodies, or indeed from the generative processes of all living bodies that make donation possible, are not viewed as generating value in the same way as the labour expended by the scientist to transform these bodily materials into a scientific object, medical research or therapy (Dickenson 2007; Waldby and Mitchell 2006; Haimes and Taylor 2015). This economy of the donation system shapes the second biopolitical figure of the 'clinical labourer' discussed in the next section of the essay. Given the vitality of living bodies, transforming this vitality into (scientific) objects is at once a familiar economic transformation echoing previous forms of the commodification of living labour, but equally disorienting because it relies on obscure bodily processes, attributions of ownership through patents on 'life', technical transformations of the body's molecular vitality, and the extension into new domains of gendered forms of exploitation.

THE PERFORMATIVE FIGURE OF THE 'CLINICAL LABOURER'

Early critical perspectives on reproductive technologies and the use of bodily tissues arising from fertility therapy, pregnancy and birth (e.g., eggs, cord blood, etc.), tended to identify dissonance between patients' and donors' intuitive claims to ownership of their bodies and the juridical prohibition against ownership of one's own or another's body. Arguments in favour of granting property rights to donors of human tissues, including by feminist scholars, were made in the context of the preeminent role of intellectual property as the primary means by which to generate surplus in the knowledge economy (Dickenson 2007). Existing legal regimes of property recognise the transformation of a body part into a research tool, but not the claims of the originator or donor of this part to ownership of the original material. As feminist scholars have noted, much of the literature on biomedical technologies tends to focus on the 'cognitive or knowledge labour' of the new biotech industries, rather than on the tissues and cells that provide the resources or materials for research (Waldby and Cooper 2008, 9). In contrast to the liberal presumption of the relations between researchers and donors/patients as oriented around the transfer of body parts as 'property in the body', other feminist readings of the production of value in the bioeconomy articulate the necessity to theorise women's bodily participation as egg donors, experimental subjects, surrogates and contributors of research tissues as a form of 'clinical labour' (Cooper and Waldby 2014).

This perspective aligns novel forms of embodied labour to other historical forms of reproductive or sexualised work performed by women and invites further consideration of how the sphere of social reproduction, broadly conceived as the daily and generational reproduction of labour, is grounded in bodily processes now being ceaselessly transformed and reworked in the donation and tissue economy. Analysis of these new forms of embodied labour could much more productively be brought together with the extensive critical and Marxist feminist literature on social reproduction. This call is increasingly echoed by theorists of social reproduction who engage with 'new materialist' approaches to the body that emphasise the re-emergence of bodily or biological processes as central concerns of feminist analysis (Meehan and Strauss 2015). The conceptualisation of new forms of embodied, clinical labour further confounds the distinction between the spheres of productive and reproductive labour critiqued by feminist theorists of social reproduction (Mitchell, Marston and Katz 2003) and points to the new ways in which the extraction of value is reliant on the body's metabolic and hormonal rhythms (Cooper and Waldby 2014).

The now-global markets for oöcytes, or female germ cells that can become fertilisable eggs, provides an apt example. In the United States, the sale of oöcytes is virtually unregulated in contrast to many other jurisdictions, including Canada, New Zealand, Australia, Singapore, and most of Western Europe, where stringent regulations govern the exchange of oöcytes and a steady supply is unavailable. Melinda Cooper and Catherine Waldby argue that new forms of globalised reproductive labour can thus be tracked from the oöcyte markets of Eastern Europe to the clinical recruitment sites of China and India that supply the raw materials for assisted reproduction and research into regenerative medicine and therapeutic cloning. Eggs are now exchanged in a global human egg market, the production of which they argue represents a 'very literal form of bodily, reproductive labour' (Waldby and Cooper 2008, 8). Women's 'clinical labour' is now harnessed less for efforts at nation-state building through a biopolitics of population, and instead channelled into private or joint private-public investments in the biotechnology industry and its clinical application.[1]

Waldby and Cooper suggest that part of women's contributions to the rapid growth of the biotechnology sector can be tracked in two ways: through the networks of tissue extraction and circulation resulting from egg donation or sale for use in assisted reproduction technology treatments; and in the importance of stem cells derived from 'reproductive tissues (embryos, oöcytes, cord blood)' (2008, 3) to regenerative medicine and other medical applications of biotechnology research. Summarising this new site of women's bodily labour, Waldby and Cooper argue that these emergent forms of reproductive labour are more closely associated with the entrepreneurialism of private biotechnology firms, even as those firms may also work with state governments and are subject to the vagaries of national regulatory climates. Waldby and Cooper draw evocative links between previous forms of primitive accumulation that relied on women's reproductive labour and the new markets for eggs and other reproductive tissues that have yet to be fully mapped out in the literature on biomedical research. As a form of labour, participation in the 'bioeconomy' is stratified, both along lines of privilege, as well as between women who donate 'surplus' eggs and embryos from IVF treatment and women who sell their reproductive tissues. Clinical labour confounds analyses that take either the industrial worker or the immaterial-intellectual labour of the scientist as a model for new forms of bodily extraction at work in the contemporary bioeconomy. Furthermore, clinical labour as a concept is a performative rendering of the possibility for a new subject position for the surrogate, tissue donor, or clinical trial participant, and opens up conceptual avenues obscured by the emphasis on liberal claims to a right of 'property in the body'.

Clinical labour thus offers analytical insights into the biopolitics of women's bodies in the highly technologically mediated world of contemporary reproduction, connecting participation in clinical trials to the creation of global egg markets, transnational surrogacy, and the growth of personal biobanking. Clinical labourers make bodily contributions to the bioeconomy by providing eggs, tissues or access to their body's vital processes, but the benefits of associated research studies do not necessarily accrue to them. Clinical labour also suggests new possibilities for political mobilisation by donors, surrogates and trial participants *as* labourers, underscoring how transformations to the very definition of the labourer as subject are central of the growth of institutions geared towards gaining access to the life processes of bodies. Volunteers in drug trials in the United States, for example, draw on entrepreneurial modes of self-cultivation—as well as means of avoiding compliance and 'gaming' the trial system—in order to navigate the conditions of a precarious labour market and economy (Monahan and Fisher 2015).

By contrast, research on gestational surrogacy in India suggests that surrogates draw on alternative values and meanings of surrogacy as 'different from a job', countering the purely transactional logics of contractual labour in which they participate by emphasising the 'transcendental' nature of their gift. At the same time, surrogates also pragmatically pursue ways to extend the relationship with the intending parents for their own or their families' future material benefit (Vora 2015, S103). The mobilisation of the figure of the clinical labourer may therefore be complicit with modes of biopolitical governance but also enable the exercise of resistance to exploitation and precarity (through 'gaming' the system). In other ways, resistance could also mean resistance to the definition of one's labour as purely transactional, drawing on the counternarrative of surrogacy as 'more than a job' in order to more effectively mobilise further financial commitments on the part of the contractor/employer.

However, there is no agreement over the extent to which embodied, clinical labour is seen as central to the 'real' economy of the biosciences, biotech, big pharma and contemporary medicine and health care. In a recent analysis of the bioeconomy, 'science and technology studies' (STS) approaches are positioned in opposition to those of 'political economy' in a call to replace critical and feminist STS imaginaries of the (bio)political economy with a 'real' notion of economic value. The authors argue that 'value', as surplus value, is actually only produced through the protected domain of intellectual property rights, in which the agents of value are none other than the scientists who can claim those rights (Birch and Tyfield 2012). The knowledge economy, from this perspective, is made ever more immaterial, obscuring the embodied nature of the work of knowledge production itself, and even further the bodily contributions that form the resources on which such immaterial production depends.

What is striking about the staging of this debate and the reluctance to 'embody' the bioeconomy is the extent to which something like a theory of 'clinical labour' is already at work. Viewing clinical trial participants as workers rather than volunteers may actually align more closely with how participants view their contribution. Roberto Abadie describes a group of professional 'guinea pigs' who view their participation in clinical trials as a different kind of work akin to what one participant called the 'mild torture economy, you are not being asked to produce or to do something anymore, you are being asked to endure something' (2010, 2). Trial participants, some with union-organising backgrounds and experience, successfully organised a walkout to protest against compensation rates and the changing schedule of trial activities after the initial trial contract had been signed (see also Monahan and Fisher 2015). Clinical trial participants may not experience payment as compensation for a gift but instead regard clinical trial participation as another form of precarious work, despite what bioethical norms of recruitment suggest. With regard to surrogacy as a form of labour, legal scholarship on global trade is exploring whether the outsourcing of pregnancy to overseas surrogates would need to be considered under the WTO's Trade-Related Aspects of Intellectual Property Rights (TRIPS) agreement on trade and remuneration in services (Stephenson 2009). In this regard, trial participation and surrogacy are already regarded as forms of clinical labour rather than as instances of 'property in the body'. This suggests that the performative calling-into-being of the proletariat of nineteenth-century capitalism has its twenty-first-century counterpart in the new forms of embodied labour that entail access to the vital processes of the living being. The outsourcing of pregnancy, as Cooper and Waldby (2014) also suggest, recalls previous configurations of the racialised outsourcing of reproductive labour under Western imperialism. They draw parallels between the dynamics of production and reproduction at work in transatlantic slave economies, wherein slave women were viewed as both productive and reproductive labourers of slave capital, and the contemporary outsourcing of pregnancy to racialised women in the Global South. Indeed, imaginaries of the surrogate womb as an 'empty' host for the child of intending parents that gains value through pregnancy and birth also recalls the colonial framing of territory as unpopulated and unproductive until brought into productivity by the coloniser's technological expertise and capital (Vora 2013, 100).

GOVERNMENTALITY AND MATERNAL SUBJECTS

Population-based biopolitics remains a relevant consideration in the contemporary bioeconomy, from the role of reproductive technologies in harnessing reproduction for colonial and nation-state building projects (Nahman 2013) to the expansion of population-based policies of 'optimization' that entail investments in human capital as a collective as well as individual project (Greenhalgh 2007). Globally, the administration of reproduction is spatially uneven and heterogeneous: continuing efforts to target the fertility of populations are concurrent with the increasing investment of maternity as an aspiration corresponding to demands for individual self-fulfilment (Murphy 2012). Work on the modalities of contemporary reproduction call attention to how the individualisation of reproduction is rendered through governmental rationales as well as forms of discipline. These new mechanisms of governmental power include the expansion of modes of 'professionally taught self-determination' (Samerski 2009) and the exercise of 'control through narrative' (Memmi 2016) in which women are encouraged by medical professionals to clearly articulate their reproductive desires and choices. In this sense, analysis of the gendered biopolitics of donation also requires attention to liberal forms of freedom to choose, to desire new forms of bodily liberation and to new governmentalities at work.

The biopolitics of donation is implicated in the shaping of maternal conduct around the bodily materials produced during pregnancy. Parents are encouraged by commercial enterprises to 'value' materials previously discarded and to retain these materials (e.g., umbilical cord blood, placenta, baby teeth) in the event they may become useful for treating future conditions suffered by the child or by other family members. As Jennie Haw (2016) writes in her study of private umbilical cord blood banking among mothers in Canada, retaining such materials in a commercial biobank requires considerable work on the part of mothers who must coordinate the various institutional and technical regulations required to successfully store a blood sample. The coordination of health-care professionals in this process is necessary but the responsibility for ensuring that samples are properly collected falls to the mother. This assumption of new responsibilities exemplifies, following Sharon Hays (1996), the performance of 'intensive motherhood' in which mothers are exhorted to assume responsibility for all aspects of the health and well-being of their children. Yet the 'double uncertainty' surrounding cord blood use—the uncertainty that it will ever be needed in the future, and the uncertainty over whether a particular cord blood sample will be sufficient if it is needed—means that mothers banking cord blood must assume the burden of these future-oriented, promissory technologies (Fannin 2013).

In the South Korean context, Yeonbo Jeong (2014) cites how private cord blood banking responds to expectations that mothers will practice a particular form of intensive parenting characterised as 'scientific' or 'professional' motherhood. Jeong writes: 'the discourse of scientific motherhood has become embedded as a "technical rationality" involving modern principles of hygiene, discipline and education. Today, "good mothers" are expected to develop expert knowledge in these and many other areas related to raising children' (364). The production of 'good mothers' who assume personal responsibility for the uncertainties of their fertility, pregnancy, birth and their child's future health are hallmarks of the governmentalising of contemporary reproduction (Fannin 2011; 2013).

The shaping of maternal conduct around the donation of materials derived from pregnancy has led feminist scholars to call for new analytical frames and concepts, from new 'modes of reproduction' appropriate to the contemporary political economies of assisted reproduction (Thompson 2005) to 'new materialist' renderings of the importance of biological processes in accounts of feminist theories of the body (Hird 2007, Neimanis 2014). The ultimate fate of tissues derived from pregnancy is increasingly part of the constellation of forces that seek to shape maternal conduct and to engender new kinds of maternal subjects, who are faced with the responsibility of considering the promissory therapeutic value of tissues created during pregnancy as well as the potential transformation of these materials into scientific objects.

BIOPOLITICS OF THE BODY: CONCLUDING REMARKS

Feminist approaches to the biopolitical dimensions of new technologies in the life sciences emphasise the need for new conceptual vocabularies to capture how bodily capacities for reproduction, growth and development are the target and object of biopolitics. The contemporary terrain of reproductive biopolitics thus involves new forms of bodily contribution and new configurations of living labour, and calls into being new subjectivities and new social relations. Work on the biopolitics of reproduction highlights how disciplinary control and governmental self-fashioning shape the uses of human tissue in the bioeconomy. This chapter explores gendered aspects of tissue donation in light of the 'procreative hinge' that Penelope Deutscher (2012) identifies as an important yet hitherto overlooked dimension of biopolitics. I emphasise how donation economies of the life sciences value the 'biological' differences of sexed and gendered bodies in specific ways.

Gender, labour and reproduction continue to be contested terms in a Foucauldian analytic but this chapter argues for the significance of integrating the reproduction—as both a problem of population as well as governmental self-fashioning—into the frame of biopolitics. These overlapping modes of biopolitical practice are the focus of critical and feminist accounts of the use, circulation and exchange of bodily materials and of the conduct of maternal subjects of donation. The biopolitics of population targets sexed bodies as sites for the administration of collective growth and health, an observation deeply resonant with interpretations of Foucault's writings on biopolitics. At the same time, the transformation of motherhood into an individualised aspiration bound up with practices of self-actualisation and entrepreneurial orientations towards the future health of one's child underscores the historical transformation of human reproduction from the domain of 'natural history' to 'biology' to 'psychology', and the emergence of new governmentalities of pregnancy and birth.

NOTE

1. State-centred population biopolitics and biopolitical investments in an individual's access to reproductive technologies can converge in the articulation of recurring anxieties over declining birth rates in Western Europe and the viability of social welfare systems. For example, in 2006, British media discussed the growth of IVF as a 'solution to population woes' in the United Kingdom, lamenting that although IVF was 'good for the British economy', the British government had yet to develop a comprehensive population policy promoting IVF to complement its attention to the issue of immigration (Ryan 2006; BBC 2006).

BIBLIOGRAPHY

Abadie, Roberto. 2010. *The Professional Guinea Pig: Big Pharma and the Risky World of Human Subjects*. Durham, NC: Duke University Press.

Allen, Amy. 2013. "Feminism, Foucault, and the Critique of Reason: Re-reading the History of Madness." *Foucault Studies* 16: 15–31.

Almeling, Rene. 2009. "Gender and the Value of Bodily Goods: Commodification in Egg and Sperm Donation." *Law and Contemporary Problems* 72: 37–58.

Almeling, Rene. 2011. *Sex Cells: The Medical Market for Eggs and Sperm*. Berkeley: University of California Press.

Bashford, Alison. 2014. *Global Population: History, Geopolitics, and Life on Earth*. New York: Columbia University Press.

BBC. 2006, June 30. "IVF Could 'Ease Population Woes.'" news.bbc.co.uk/go/pr/fr/-/1/hi/health/5098324.stm. Accessed 1 March 2017.

Bennett, Jeffrey A. 2009. *Banning Queer Blood: Rhetorics of Citizenship, Contagion, and Resistance*. Tuscaloosa: University of Alabama Press.

Birch, Kean. 2012. "Knowledge, Place, and Power: Geographies of Value in the Bioeconomy." *New Genetics and Society* 31 (2): 183–201.

Birch, Kean, and David Tyfield. 2012. "Theorizing the Bioeconomy: Biovalue, Biocapital, Bioeconomics or . . . What?" *Science, Technology & Human Values* 38 (3): 299–327.

Brown, Nik, and Alison Kraft. 2006. "Blood Ties: Banking the Stem Cell Promise." *Technology Analysis and Strategic Management* 18 (3/4): 313–27.

Cooper, Melinda, and Catherine Waldby. 2014. *Clinical Labor: Tissue Donors and Research Subjects in the Global Bioeconomy*. Durham, NC: Duke University Press.

Deutscher, Penelope. 2008. "The Inversion of Exceptionality: Foucault, Agamben, and 'Reproductive Rights.'" *South Atlantic Quarterly* 107 (1): 55–70.

Deutscher, Penelope. 2010. "Reproductive Politics, Biopolitics and Auto-Immunity: From Foucault to Esposito." *Journal of Bioethical Inquiry* 7 (2): 217–26.

Deutscher, Penelope. 2012. "Foucault's History of Sexuality, Volume I: Re-reading Its Reproduction." *Theory, Culture & Society* 29 (1): 119–37.

Dickenson, Donna. 2007. *Property in the Body: Feminist Perspectives*. Cambridge: Cambridge University Press.

Fannin, Maria. 2011. "Personal Stem Cell Banking and the Problem with Property." *Social & Cultural Geography* 12 (4): 339–56.

Fannin, Maria. 2013. "The Burden of Choosing Wisely: Biopolitics at the Beginning of Life." *Gender, Place & Culture* 20 (3): 273–89.

Fannin, Maria, and Julie Kent. 2014. "Origin Stories from a Regional Placenta Tissue Collection." *New Genetics and Society* 34 (1): 25–51.

Foucault, Michel, and Arlette Farge. 2016. *Disorderly Families: Infamous Letters from the Bastille*. Minneapolis: University of Minnesota Press.

Goven, Joanna, and Vincenzo Pavone. 2014. "The Bioeconomy as Political Project: A Polanyian Analysis." *Science, Technology, & Human Values* 40 (3): 302–37.

Greenhalgh, Susan. 2007. "The Chinese Biopolitical: Facing the Twenty-First Century." *New Genetics and Society* 28 (3): 205–222.

Haimes, Erica, and Ken Taylor. 2015. "Rendered Invisible? The Absent Presence of Egg Providers in the U.K. Debates on the Acceptability of Research and Therapy for Mitochondrial Disease." *Monash Bioethics Review* 33: 360–78.

Haw, Jennie. 2016. "Corporeal Commodification and Women's Work: Feminist Analysis of Private Umbilical Cord Blood Banking." *Body and Society* 22 (3): 31–53.

Hays, Sharon. 1996. *The Cultural Contradictions of Motherhood.* New Haven, CT: Yale University Press.

Helmreich, Stefan. 2008. "Species of Biocapital." *Science as Culture* 17 (4): 463–78.

Hird, Myra. 2007. "The Corporeal Generosity of Maternity." *Body & Society* 13 (1): 1–20.

Jeong, Yeonbo. 2014. "Scientific Motherhood, Responsibility, and Hope: Umbilical Cord Blood Banking in South Korea." *New Genetics and Society* 33 (4): 349–69.

Kent, Julie. 2008. "The Foetal Tissue Economy: From the Abortion Clinic to the Stem Cell Laboratory." *Social Science & Medicine* 67 (11): 1747–56.

Kent, Julie, Maria Fannin and Sally Dowling. 2017. "Gender Dynamics in the Donation Field: Human Tissue Donation for Research, Therapy and Feeding." Unpublished essay.

Kent, Julie, and Anne-Maree Farrell. 2014. "Risky Bodies in the Plasma Bioeconomy: A Feminist Analysis." *Body & Society* 21 (1): 29–57.

Larsen, Lars Thorup. 2007. "Speaking Truth to Biopower." *Distinktion: Journal of Social Theory* 8 (1): 9–24.

Martucci, Jessica. 2010. "Negotiating Exclusion: MSM, Identity, and Blood Policy in the Age of AIDS." *Social Studies of Science* 40 (2): 215–41.

McNay, Lois. 2009. "Self as Enterprise: Dilemmas of Control and Resistance in Foucault's The Birth of Biopolitics." *Theory, Culture and Society* 26 (6): 55–77.

Meehan, K., and K. Strauss, eds. 2015. *Precarious Worlds: Contested Geographies of Social Reproduction.* Athens: University of Georgia Press.

Memmi, Dominique. 2016. "Une Discrète Naturalisation de la Maternité : Le For Intérieur Féminin Face aux Aléas de la Feproduction" ["A Discreet Naturalization of Motherhood: Feminine Subjectivity in the Face of the Hazards of Reproduction"]. *Sociologie* 4 (7). http://sociologie.revues.org/2957. Accessed 12 August 2017.

Mills, Catherine. 2017. "Biopolitics and Human Reproduction." In *The Routledge Handbook of Biopolitics*, edited by Sergei Prozorov and Simona Rentea, 281–94. Abingdon: Routledge.

Mitchell, Katharyne, Sallie A. Marston and Cindi Katz. 2003. "Introduction: Life's Work: An Introduction, Review and Critique." *Antipode* 35 (3): 415–42.

Monahan, Torin, and Jill A. Fisher. 2015. "'I'm Still a Hustler': Entrepreneurial Responses to Precarity by Participants in Phase I Clinical Trials." *Economy and Society* 44 (4): 545–66.

Murphy, Michelle. 2012. *Seizing the Means of Reproduction: Entanglements of Feminism, Health and Technoscience.* Durham, NC: Duke University Press.

Nahman, Michal. 2013. *Extractions: An Ethnography of Reproductive Tourism.* London: Palgrave Macmillan.

Neimanis, A. 2014. "Speculative Reproduction: Biotechnologies and Ecologies in Thick Time." *philoSOPHIA* 4 (1): 108–128.

Parry, Bronwyn, Beth Greenhough, Tim Brown and Isabel Dyck, eds. 2015. *Bodies across Borders: The Global Circulation of Body Parts, Medical Tourists and Professionals.* Abingdon: Routledge.

Petchetsky, Rosalind Pollack. 1987. "Fetal Images: The Power of Visual Culture in the Politics of Reproduction." *Feminist Studies* 13 (2): 263–92.

Petersen, Alan. 1996. "Risk and the Regulated Self: The Discourse of Health Promotion as Politics of Uncertainty." *Australia and New Zealand Journal of Sociology* 31 (1): 44–57.

Pfeffer, Naomi. 2009. "How Work Reconfigures an 'Unwanted' Pregnancy into the 'Right Tool for the Job.'" *Sociology of Health & Illness* 31: 98–111.

Rajan, Kaushik S. 2006. *Biocapital: The Constitution of Postgenomic Life*. Durham, NC: Duke University Press.

Repo, Jemima. 2015. *The Biopolitics of Gender*. Oxford: Oxford University Press.

Richardson, Angelique. 2003. *Love and Eugenics in the Late Nineteenth Century: Rational Reproduction and the New Woman*. Oxford: Oxford University Press.

Rose, Nikolas. 2007. *The Politics of Life Itself: Biomedicine, Subjectivity and Power in the Twenty-First Century*. Princeton, NJ: Princeton University Press.

Rose, Nikolas, and Carlos Novas. 2004. *Biological Citizenship*. London: Blackwell.

Ryan, Caroline. 2006, June 19. "IVF 'Good for British Economy.'" news.bbc.co.uk/go/pr/fr/-/1/hi/health/5095884.stm. Accessed 14 August 2017.

Samerski, Silja. 2009. "Genetic Counseling and the Fiction of Choice: Taught Self-Determination as a New Technology of Social Engineering." *Signs* 34 (4): 735–762.

Santoro, Pablo. 2011. "Liminal Biopolitics: Towards a Political Anthropology of the Umbilical Cord and the Placenta." *Body & Society* 17 (1): 73–93.

Shildrick, Margrit. 1997. *Leaky Bodies and Boundaries: Feminism, Postmodernism and (Bio) Ethics*. New York: Routledge.

Steinman, Judith L. 2006. "Gender Disparity in Organ Donation." *Gender Medicine* 3 (4): 246–52.

Stephenson, Christina. 2009. "Reproductive Outsourcing to India: WTO Obligations in the Absence of US National Legislation." *Journal of World Trade* 43: 189–208.

Stoler, Ann Laura. 1995. *Race and the Education of Desire*. Durham, NC: Duke University Press.

Strong, Thomas. 2009. "Vital Publics of Pure Blood." *Body & Society* 15 (2): 169–91.

Thompson, Charis. 2005. *Making Parents: The Ontological Choreography of Reproductive Technologies*. Cambridge, MA: MIT Press.

Titmuss, Richard. 1970. *The Gift Relationship: From Human Blood to Social Policy*. London: Allen & Unwin.

Vora, Kalindi. 2015. *Life Support: Biocapital and the New History of Outsourced Labor*. Minneapolis: University of Minnesota Press.

Waldby, Catherine, and Melinda Cooper. 2008. "The Biopolitics of Reproduction: Post-Fordist Biotechnology and Women's Clinical Labour." *Australian Feminist Studies* 23: 57–73.

Waldby, Catherine, and Melinda Cooper. 2014. *Clinical Labor: Tissue Donors and Research Subjects in the Global Bioeconomy*. Durham, NC: Duke University Press.

Waldby, Catherine, and Robert Mitchell. 2006. *Tissue Economies: Blood, Organs, and Cell Lines in Late Capitalism*. Durham, NC: Duke University Press.

Weir, Lorna. 2006. *Pregnancy, Risk and Biopolitics: On the Threshold of the Living Subject*. London: Routledge.

Chapter 7

Mapping the Will for Otherwise

Towards an Intersectional Critique of the Biopolitical System of Neoliberal Governmentality

Charlie Yi Zhang

On 20 January 2017, with his fiery and virulence-laced inaugural speech, the forty-fifth president of the United States, Donald J. Trump, stunned the world. Hailing 'America first' from the western front of Capitol Hill, he painted a bleak picture of the United States and cast millions of Americans as immiserated victims of globalisation. Promising that 'the American people', under his aegis, would 'be forgotten no longer', Trump threw into relief a nationalist vision for remaking the country.[1] Just three days earlier, Chinese president Xi Jinping, in his plenary speech at the World Economic Forum meeting at Davos, took a different stance of globalisation and fervently defended its benefits with cosmopolitan sensibilities. China, as Xi posits, is ready to take up the leadership abandoned by Trump's United States to push forth the globalising process, marking a sharp departure from the nationalist wave that is engulfing the Western world.[2]

Despite their contradicting worldviews and tones, Trump and Xi nonetheless converge towards an identity-paved and biopolitically formulated road that they lay out to lead their visionaries into fruition. Riding a racist, xenophobic, misogynist and ableist campaign into the White House, Trump vows to disrupt the identity-grounded 'political correctness', upend the political establishments, and bring jobs back to help the suffering workers to make 'America great again'.[3] Comparably, in the policy package that Xi proposed in 2013 to initiate the most far-reaching socioeconomic reforms since 1978, the plan to reshape the Chinese population by reforming the one-child policy and household registration system *Hukou* attracted the most attention.[4] These policies, as I have argued elsewhere (Zhang 2014a), building on the biopolitics of gender and class, have recreated tangible/legalistic and intangible/cultural boundaries to produce a huge segment of flexible and exploitable population in urban space that energises and sustains China's marketising thrust and reintegration with the global economy.

As has been made clear recently, the identity-paved road along the competing nationalist/globalist directions, as laid out by Trump and Xi, respectively, would actually lead to the same neoliberalising goal. As the biopolitics of gender and class played a vital role in creating a population dividend that bolstered China's neoliberalisation launched in the late 1970s (Zhang 2014b; 2016), Xi's administration is gearing up to reshape the Chinese population in line with broad-scale structural changes to safeguard the Chinese economic juggernaut into the new global leader. In the meantime, Trump's promise to salvage the working poor is debunked by the 'richest cabinet' that he put together as well as the privatising, deregulating and welfare-slashing neoliberal policies that he intends to put into effect.[5] In a nutshell, different nationalist/globalist visionaries, along the similarly identity-cemented road, work towards the same neoliberalising *telos*.

Regarding these rivaling realities concatenated through baffling and specious political narratives and tactics, Foucault provides us with instructive tools for deciphering and apprehending. As a set of flexible and contingent governing practices—or what he calls 'governmentality'—to valorise and perpetuate self-serving market fundamentalism as the paramount principle for reconstituting human life and societies, neoliberalism has steered and stewarded global restructuring through interpenetrating deployment and diagonal exertion of disciplinary and sovereign power (Foucault 2008). Centring on singular bodies, the former labours to individualise subjects as self-interest-driven entities to *live* the marketising telos (Foucault 2003; 2008). Targeting the collective body 'population', the latter permeates, envelopes, subtends and perpetuates social institutions and structures—both discursive and material—to create conditions and delimitations for *the living of* the marketising telos by the individualised subjects (Foucault 2003; 2008). Indeed, as a specific form of governmentality, the raison d'être of neoliberalism lies in the production and inculcation of what neoclassic economists call 'Homo Economicus'—a rational human subject motivated for maximising personal interests—to induce and sustain market-engineered societies (Foucault 2008).

Using sexuality as an example, Foucault (1978) further specifies an identity-nexused interface of the transversal workings of biopower and biopolitics in line with broader politico-economic systems, yet leaves room for further articulation. The burgeoning of the capitalist system, he contends, garnered fundamental momentums from the sexualised disciplinary practices of individual bodies and regulative attempts over populations in eighteenth- to nineteenth-century Europe. Through discursive formations, scientific researches and medical practices, and with support by juridical protocols, legal practices and social institutions of state, family and school, homosexuality was created as a categorical antithesis to throw light on an obscure and nebulous concept of 'heterosexuality'. Meanwhile, via criss-crossing maneuvers of biopower and biopolitics, other 'abnormal' sexual categories, such as child sex, adolescent masturbation, and extramarital relationships were produced and circulated along with the demonised 'homosexuality' to constrict and delimit the human sexual spectrum in ways to valorise the heterosexual conjugal relationship as the normative way of being. Driving sexuality as a wedge-apparatus between macromanagement of the populations and micropolitics of self-regimentation, the normalised selfhood serves to channel people's heterogeneous libidinal energies and desiring fantasies into the monolithic purpose of reproducing a quality body, and consolidate the nuclear family as the basis for inculcating subjects that best suit capital's needs. In contrast to his well-substantiated analysis of the sexual apparatus of social management and individual disciplining, Foucault (2003), however, takes a different approach to race, and uses it interchangeably with 'racism' for deductive explanations of the role that race/racism, since the nineteenth century, has taken in producing a 'pure' population to police national boundaries and fortify state power. This reductive take not just runs counter to his signature genealogical approach but, as Ann Laura Stoler holds, also entails 'a clean erasure of the question of racism from his project' (1995, 25).

Feminist scholars, building on critiques of Foucault, not only carve out new directions and dimensions that he never broached—such as gender, race, ethnicity and dis/abilities, to name a few—but extend the scope of the identity-grounded governance from its Euro-American bastion to colonies. For instance, as Rey Chow posits, 'seen in the light of biopower, sexuality is no longer clearly distinguishable from the entire problematic of the reproduction of human life that is, in modern times, always racially and ethnically inflected' (2002, 7). Centering on Foucault's lacuna, Chow and Stoler unpack how the racialised/ethnicised regulation of bodies and lives is played out for the making of both colonising and colonised subjects to spread the capitalist system across the world. Situating the formulation and reformulation of biopower and biopolitics at the core of the day-to-day running of the imperial and colonial systems, these scholarly endeavors have broadened our horizon to probe neoliberal globalisation as the recent revamping of these systems.

With that said, highly informative as Foucault's instructions are, questions still remain. In particular, instead of his sexuality-focused, one-layered and flattened analytics, how do other identity categories, such as race, class and gender, interconnect and overlap with sexuality to shore up the governing system of global neoliberalism? Given the current mechanism and arrangement spawned by neoliberal governmentality on a global scale, how can new forms of livelihood and socialities emerge, survive and thrive? And, if all potential life is embodied and lived on a daily basis, as Foucault suggests, how can scholars committed to social justice and change uncover and access these potentialities for a more humanitarian futurity? As Elizabeth Povinelli argues, we 'should look not at what he [Foucault] solved but at what remains open in his wake' (2012, 453). As I see it, Foucault might not provide us with direct answers to these questions, yet offers an instrument that he calls 'the ontology of self' (2011) to help forward our engagement with these crucial questions. Indeed, if our own individuality is the product of the intensified relationship of multifarious forms of disciplinary and sovereign power, our bodily and embedded experience would be a prime optic to unravel these complicated and entangled relationships.

Centring on the author's bodily and sociopsychological dis/relocation along a cross-China and transpacific odyssey as the analytical minefield, this chapter aims to map the mechanism that subtends and upholds global neoliberal governmentality. First, building upon feminist/queer analysis of colour, it extends Foucault's single-axis framework of sexuality to explore how gender, sexuality, class and race intersect to create a cross-border network of power at the core of neoliberal governmentality. Second, drawing upon Lauren Berlant's work, it extends the Foucauldian conception of neoliberal governmentality that functions as 'a world-homogenising sovereign with coherent intentions that produces subjects who serve its interests' (2011, 15) and illustrates how people's affective attachments to an array of fantasies of a good life defined by meritocratic class mobility, gendered bourgeois normativity and heteronormative intimacy are created and resuscitated to engineer and sustain global neoliberalism. As I argue, the promises offered by good-life fantasies play a vital role in the production of myriad domestic and transnational Chinese migrants willing to work for the interests of capital, often against their rational calculation and well-being. Finally, following the phenomenological avenue that Sara Amhed (2007) carves out, it explicates how the noticeability of the 'arrival' of my own body in a racially charged new world that my 'Asianised' body is unable to fit in raises new consciousness of resistance. Taken together, the reflexive analysis of my physically and sociohistorically situated bodily experience informs my new critical understanding of global neoliberal governmentality, which, hopefully, will serve as a vantage point for building collective will for otherwise.

SELF-FORMATION IN CHINA'S WHIRLWIND CHANGES

I was born in a working-class family in a small town in southwest inland China in 1979, one year after Deng Xiaoping announced the grandiose economic reform and opening-up plan as a response to mounting sociopolitical and economic pressures. Abandoning Mao's stringent class-struggle policies that had entailed massive tumult and crisis during 'the Cultural Revolution', the state introduced neoliberalism as what Aihwa Ong (2006) calls 'exception' into rural areas to experiment with the market mechanism. 'Crossing the river by feeling the stones', Deng, the 'general architect of China's reforms', admonished. With that said, since 1978, allowing farmers in certain areas to sell their extra produce in the market after turning in tax grains, the state accumulated the experience about how to run and manage the society through market rather than planning, and then gradually expanded the practice to coastal and east/south urban regions in the 1980s, thus avoiding a Soviet-style politico-economic overhaul and maintaining a smooth transition. It should be noted that, partially designed by the state, partially experimental and exploratory, China's neoliberalising thrust is also grounded in the coordination between measured attempts to test the water of the market and the calculated management of its own population (Zhang 2014a). In early 1979, the one-child policy was introduced by the state to 'enhance the population quality' as part of China's modernising initiative, and one year after, the eugenically inflected rhetoric of 'the planned family' was written into the Constitution as a fundamental national policy.

Despite the absence of oftentimes warming and cheering companionship by siblings, the memory of my early years is nevertheless quite jubilant, joyful and mostly worry-free. My celebrated childhood, not to a small extent, should be attributed to my working-class background defaulted as the norm of the time when China's bits-and-pockets neoliberalisation was still edging its way, slowly and carefully, along China's coastal line and across the southern and eastern urban regions. The devastating effects of polarisation induced by neoliberal economics, in this light, were not (yet) fully felt in my home/town in a corner of southwest hinterland China. To wit, my class identity endowed me with an array of privileges, such as the nearly free high-quality public education, affordable rationed food and medical care by virtue of my urban *Hukou* status, and the free housing as the welfare associated with my parents' tenure-job in state-owned enterprises (SOEs). Essential to Socialist China, *Hukou*, by a string of skewed welfare provisions allocated for urbanites only, imposed enormous class inequalities and created a polarised urban-rural society (Whyte 2010). Lived as a background to experience the world organised as such, my urbanised working-classed body produced through the state policies and designations enabled me to fit into a lifeworld built upon and made possible by this structure, so effortlessly and seamlessly that it did not even register in my young mind.

Besides my class background, my embodied masculinity was another source of my happy childhood memory and taken-for-granted privileges. At school, my arduous pursuit of knowledge and excellent academic performance invited frequent complimentary comments from my peers and teachers. At home, my lanky and fragile figure—largely due to a lack of proper exercise—placed me in an advantaged position versus my 'coarse, wild and ill-educated' cousins from the rural regions, further reinforcing my urbanity-granted class privileges. Indeed, different from the Western norm of hypermasculinity (Kimmel 1996), one culturally distinguished characteristic of the Chinese genderscape—until recently—was the soft, literary manhood that I embodied (Louie 2002; Song 2004). This norm of masculinity is rooted in the Confucian tradition that valorises male literati of the gentrified class as the central figures in the political order of the feudal state. In feudalist China, the patriarchal gender relationship was the important discursive resource to consolidate and legitimise the state system and the subject-monarch relationship (Bray 1997).

Many subjects internalised their constructed feminine role (subordinate to the 'masculine' monarch) through the performance of the softened masculinity, which turned into a cultural standard of manhood among the ruling class (Zhang 2008). As Tani Barlow (1994) notes, the traditional Chinese gender system is not as distinctively demarcated as in the Western context despite its equally, if not more, oppressive patriarchal system. In the Maoist era, the ambiguous gender line was revamped for a Socialist version of 'gender equality' largely based on a masculinist model: women were masculinised as equal to men in public sectors of employment, and they also gained more agency in the private domain of marriage (Croll 1983; Yan 2003). In its intersection with social class, gender was further articulated by the Maoist state as the revolutionary working-class and peasant masculinity to epitomise the socialist regime. Therefore, the imagery of workers operating machines became a central paragon of Mao's China. But even so, soft masculinity never completely lost its cultural values during Mao's time (Wang 2003).

Not surprisingly, gender, as a biopolitical parameter under the stringent control of the state, morphed along with its focus shifting to neoliberalism in the late 1970s, and the gentrified soft, literary masculinity that had been kept at bay was set loose again. In the early 1980s, the imagery of men of letters replaced the robust and sinewy workers/soldiers as the new caliber of Chinese masculinity. As Lisa Rofel (2007) suggests, as China moved away from socialism to neoliberalism, the educated men were celebrated as the new normative figure equipped with rationality and reasoning capacities, who could not only lead and support their own family but also salvage the country traumatised by the Maoist monopoly over economy and culture. By and large, my childhood conflated with this turbulent biopolitical shift. My embodied soft, fragile masculinity that was cross-referenced and consolidated by my scholastic achievements at school served as the leverage for me to stand out from a group of students with homogenous class background at the time.

As China expanded and escalated its neoliberalising transition in the late 1980s, the Chinese genderscape was reconstituted as part of the broad biopolitical moves. Sanctioned by the state, a multitude of gender discourses mushroomed to cultivate and legitimate an essentialised notion of sexual difference through mass media. On the one hand, the discourses of gender differences started to proliferate, and female images with emphasised feminine and sexual features almost dominated mass media (Croll 1995; Evans 1997; Honig and Hershatter 1988). On the other hand, a discourse of 'remasculisation' emerged to recover men from the traumatising experience in Mao's era and reconstruct the masculine prowess of men in post-Mao China (Rofel 2007; Zheng 2009; Lu 2000; Zhang 2015). All of a sudden, images of tough guys replaced the soft, literary masculinity as the new gender norm. Even decades later, I can still remember that sun-drenched summer afternoon in the late 1980s when my parents finally agreed to take me to see the Hollywood blockbuster *First Blood*. As the lengthy camera shot zoomed in on and across Sylvester Stallone's elliptical pectoral muscles, bulging arms, and tight abdominals, my heart started to beat fast and heavy, and sweat soon soaked my shirt. For a Chinese audience which had long been indoctrinated by the state for self-sacrificial allegiance and obedience, this musculature spectacle, the militant, unapologetic and unrelenting masculinity that it symbolises, and most importantly, the uncompromising rebellious spirits flickering with every 'eye for eye' revengeful scene of Rambo brought forth a disturbing and stirring experience that took me a long time to come to terms with.

China's remasculised public culture, undoubtedly, was part and parcel of the discursive shift of gender in sync with the intensified global neoliberal restructuring in the late 1980s. As Michael Wood (2006) and Lynne Attwood (1995) both note, the popularisation of hypermasculine icons like Rambo lies at the core of the antigovernment neoliberal discourses that strove to shrink the role of the state and pave the way for deregulation and privatisation. Indeed, are there better symbols of the neoliberalist market fundamentalism than these manly and unruly heroes owing and fighting to tear apart the political establishments of state?

For the state, gender might be just one biopolitical parameter at work for its daily functioning, but for many people, it is the quintessential part of their subjectivity, habituated way of living and even self-esteem. Seen in this light, the broad-scale shifting of the gendered biopolitics was harsh and challenging to me on multiple fronts. As my soft literary masculinity had been deeply inscribed onto my daily regimentation of identity and become an essential horizon for me to see, sense and perceive the world, the drastic and dramatic gender change rendered me irrevocably out of tune with the sociocultural tide waning and waxing with China's neoliberal transition. My 'not manly enough' gender performance soon degraded into a source of shamefulness, but my optic to view the world built upon and formulated through this gender lens was not flexible enough to keep me abreast of the whirlwind changes, placing me in incessant anxiety throughout my adolescent years.

China's Westernised dimorphic gendersacpe in line with its neoliberal transition needs to be further unpacked through a queer lens. As Foucault's (1978) description of the burgeoning capitalism in Europe indicates, to subject a diverse spectrum of desiring modalities to the reproductive zealotry to satisfy the increasing demand of labour for capitalist economies, all desires outside the heterosexual nuclear family needed to be policed and disciplined. In this regard, in the nineteenth century, the sodomite, previously considered a temporary aberration, was turned into homosexual as a species, through 'a certain way of inverting the masculine and the feminine in oneself' (Foucault 1978, 43). In other words, homosexuality, as a biopolitical parameter, was not only produced as the abnormal inversion of the heterosexualised gender order but also as the boundary of heterosexuality to normalise the latter as 'compulsory and natural'. Therefore, hinged on the newly constructed identity of 'homosexuality', heteronormativity and dimorphic gender configuration co-constituted and reinforced each other to guarantee that the heterosexual familial matrix is running smoothly for the burgeoning capitalist economy. For example, as Michael Kimmel (1996) shows how, in the mid-nineteenth century, the nuclear family became an important site for the professing of gender and sexual protocols to boys in the United States. Their sexual desire needed to be properly directed (towards the opposite sex) and appropriately released (for procreation and not wasted in masturbation). For men, learning how to manage their sexual desire became a litmus test to see if they could become rational and self-disciplined enough for market competition—the crucial avenue for men to prove and fulfill their family responsibilities. In other words, through the stringent policing over their sexuality, boys were not only re-created as the rational and disciplined subject for market economy but also the gendered subjects to perpetuate the nuclear family as the bedrock of capitalism.

In 1992, as Deng initiated his southern tour to resume China's neoliberalisation halted by the Tian'anmen Square massacre in 1989, a discourse of homosexuality entered mainland China, primarily through cultural products from Hong Kong and Taiwan, that further naturalised the essentialist gender differences. As Rofel (2007) and Tze-Ian Deborah Sang (2002) both note, since the 1990s, Mao's heavy-handed punitive policies to suppress homoeroticism and homosexual conduct were gradually replaced by discourses of pathologised sexual behaviour to reassemble the Chinese population. In this regard, people became 'aware' of their sexual identity and started to reexamine themselves and others via this new biopolitical prism. My carefully concealed simmering sexual desire (for men) thus became fathomable and detectable through my already delegitimised soft masculinity, subjecting me to constant scrutiny and questioning by other curious students, or even teachers. Similar to what C. J. Pascoe (2007) has observed in her ethnographic work about a US high school, hurling homophobic slurs, such as '*er yi zi*' (faggot), '*yin yang ren*' (hermaphrodite), '*tong xing lian*' (gay/homosexual), at each other became a popular game at my middle and high school in the 1990s. For many (heterosexual) boys, it was just a rhetorical swordplay to assert their heteronormative masculinity by passing along the pathological label of gayness to others. But for me, it was a hurtful and dehumanising gendered and (hetero)sexualised biopolitical surveillance and disciplining that haunted me throughout my pubertal years. Only after the dawn of the new century did I finally come to terms with my 'perverse' sexual desire. This agitated and stressed experience later became the most important impetus for me to strive for, or maybe more accurately, trade for emancipation with personal achievement—the central argument of meritocracy.

As China strode forward with its neoliberal experiments, the idea of meritocracy and personal freedom emerged and ascended to replace the socialist fundamentals of social equality and justice. By the end of the 1990s, having been released from the collective duties through an array of reforms, such as privatisation of SOEs and marketisation of housing, people were admonished to take full responsibilities for their own well-being as the state retreated from welfare provisions to enhance its administrative efficiency. Like the stories of 'pulling yourself up by your bootstraps' prevalent in Western societies, the successful experiences of self-made economic miracles by poor farmers, such as the ten-thousand-Yuan household (*wan yuan hu*), were trumpeted and parroted by mass media to beckon people to join the bandwagon of the 'socialist market economy' by starting their own businesses. For many Chinese people, having lived under the stricture of the Maoist state and its collectivised life, this rhetoric was quite appealing as it tempered with their long-subdued individualistic ambition for financial success and proved to be an effective disciplinary technique to re-create market subjects to energise and anchor China's neoliberalisation. It also unbridled and valorised the 'jungle rule' of the market—the cutthroat 'survival of the fittest' philosophy at the core of neoliberal economics. In consequence, within a short span of thirty years, China demoted from the ranks of the most egalitarian societies during Mao's reign (at least in terms of income) to one of the most unequal in terms of the distribution of wealth and opportunities (Zhang 2014a).

For youths from a poor/working-class background like me, working hard enough to become the 'fittest' and surviving one in the competition became one of—if not the only—available avenues to lift ourselves out of the ordeal for a better life. As I recall, for many years, the constant stress and anxiety emanating from my intersecting and combined non-normative gender, sexual and class identities became the major sources foraying an 'aspiring fantasy' of 'a normative self' that unfailingly fuelled and refuelled my persistent pursuit of personal achievements, and thus ultimate freedom, and that directed me along a cross-China journey and then a transpacific odyssey to find my 'right' place in the world, and thus eternal inner peace.

As China's marketisation has restored class inequalities, gender is not only (hetero)sexualised but also middle-classed. As Tiantian Zheng (2009) argues, the 1990s has seen the prevalence of what she calls 'entrepreneurial masculinity' that centred on the control of women's bodies and sexuality, blatant consumerist capabilities and bald entrepreneurial adventurism. As mentioned earlier, when China was increasingly reintegrated into the global economy after Deng's 1992 southern tour, its genderscape also blended into the global backdrop as the Western money-emblasoned masculinity became a new norm. My inferior class background, of course, further delegitimised my gender identity for my inability to participate in the 'real man' proving consumption game. But not until the late 1990s when I left my home for college had I, for the first time, truly felt the transformative power of neoliberal economics through the intersecting parameters of class, gender and sexuality.

THE GENDERED, CLASSED AND HETEROSEXUALISED 'ASPIRING FANTASY'

In 1998, when my hinterland home/town accelerated the transformative process by the magic of neoliberal economics, the spell of the market had already been put on full display in the eastern metropolitan areas where my university is located. My escapist journey—away from my parochial small home/town, from the scrutinising eyes of my peers and teachers, from my traumatising gender, sexual and class non-normativity—turned out to be more troubling and disturbing than freeing. During the four years of college, the bedtime conversation in our dormitory was almost invariably focused on the competition over masculinity, with a roomful of just come-of-age young adults boasting of *their* own masculine prowess to win the title of 'the king of the night'. In this heterosexually charged rhetorical game, money usually made the final decision. The roommates from the affluent eastern coastal regions/capital city of Beijing kept increasing the stakes of the game by flaunting their consumerist capabilities to best each other. One night, as one of them proclaimed, he and his girlfriend(s) just spent fifty dollars in a karaoke bar overnight—an amount enough to feed me for over a month! Light as the message might sound—or was made to sound light by my proud roommate—the connotation that it conveyed was heavy. It made crystal clear that our masculine and heterosexual normativity is pivoted upon the availability of monetary resources. The lack thereof automatically deprived poor students like me of the entitlement to even participate in this game. For those of us from a marginalised class background, the spellbinding neoliberally constituted dream that 'more education will finally bring you a good life' was the only viable solution to our troubled gender, sexual and class identities, beckoning us into the self-help fantasy of normalcy by investing more time and energy to remake ourselves as such. I was the lucky one: in 2004, my hard work warranted me the admission to a top-rated graduate programme in Shanghai—a ticket to the good life that I had long been aspiring to.

A centre of the global metropolitan constellations, Shanghai, by virtue of its advantaged geopolitical location and colonial history of bourgeois legacy, has become a 'hub' that links the bountiful raw materials and labourers in interior China with transnational capital (Sassen 2009). And as the cartographic 'dragon head' of the Chinese territory, Shanghai is also embraced by the state to spearhead China's marketisation (Wu, Xu and Yeh 2007). This positioning places Shanghai in the epicenter of multifarious discursive forces unleashed by neoliberalisation. As Cheng Li (1997) notes, the zealotry for Western modernism and bourgeois lifestyle gleaned from Shanghai's wartime colonial legacy re-thrives after held at bay by the Maoist state for over three decades. As numerous people from other parts of China, particularly poor rural areas, are flooding to Shanghai for job opportunities, the ethnocentrism of Shanghai natives coupled with the colonial discourse of Western modernity is translated into the thinly veiled nativist discrimination against less-developed Chinese areas in general and migrants from these areas in particular.

As one of the myriad dream-pursuing new Shanghainese, I partially internalised the colonial legacy of Western supremacy imbricated in Shanghai local culture, but meanwhile, my aspiring fantasy was shattered by the compounded nativist marginalisation and spiraling class inequalities. Every year millions of people come to Shanghai in pursuit of a better life only to find their dream get erased by the reality: the skyrocketing housing prices, the overt discrimination and even the right to love disenfranchised by the 'jungle rule' of the market. For us, the complicated feelings are often crystalised into a polarised perspective of the anticommunist outrage on the one hand, and apologetic expectations of emancipation in/by the West/United States on the other. As Stanley Rosen (2003) notes, in contemporary China, urban Chinese youth have developed two disparate images of the United States: a highly negative view of American hegemonism abroad alongside a highly positive assessment of American values and lifestyles at home. Like millions of young people who are antagonistic to the party-state but never have a chance to live in the United States, the projection of the aspiring fantasy of the good life onto the other side of the Pacific and the dream-shattering alienation from the Shanghai urban life blended into a polarised sensory/cognitive interface for me.

Regarding the conundrum that millions of good-life-pursuers encounter in Shanghai—the epicenter of China's marketisating and globalising endeavors—Berlant (2011) has coined the term 'cruel optimism' for expounding. As she frames, optimism is 'the force that moves you out of yourself and into the world in order to bring closer the satisfying something that you cannot generate on your own but sense in the wake of a person, a way of life, an object, project, concept, or scene' (2011, 1–2). And 'one of optimism's ordinary pleasures is to induce conventionality, that place where appetites find a shape in the predictable comforts of the good-life genres that a person or a world has seen fit to formulate' (Berlant 2011, 2). In the meantime, optimism can also be cruel 'insofar as the very pleasure of being inside a relation have become sustaining regardless of the content of the relation, such that a person or a world finds itself bound to a situation of profound threat that is, at the same time, profoundly confirming' (Berlant 2011, 2). In this regard, neoliberalism is not just a homogenising sovereign built upon coherent intentions to produce subjects who serve its interests, 'such that their singular actions only seem personal, effective, and freely intentional' (Berlant 2011, 15), as suggested by the Foucauldian conception of biopolitics and governmentality (2008), and more importantly, takes shapes through 'the messy dynamics of attachment, self-continuity, and the reproduction of life that are the material scenes of living on in the present' (Berlant 2011, 15).

For this sometimes self-fulfilling, sometimes self-defeating convoluted relationship, Sara Amhed (2010) proffers useful clarifications. As she puts it, things with a promise of confirming happy endings, such as what Berlant calls 'optimism', operate through future tenses and follow such logic as 'if I had . . . , I would have . . .' or 'If I should . . . , I would. . .' In other words, since the promises are perpetually projected onto future, there is no at-present ground for its 'failure'—instead of falling apart, the unfulfilled ones are but deferred, for now. Despite their futuristic orientation, the affective attachments/forces that the aspiring fantasy generates between us and the object that carries the promise are nonetheless contemporary, grounded and *real*.

Rather than simply the top-down effects of the state's biopolitically induced regulative and managerial endeavors, a set of aspirational dreams of a good life have been concretised in China, in terms similar to what Berlant (2011) has noted in Europe and the United States—a property-based heteronormative conjugal relationship blessed by meritocratically afforded social mobility. This aspiring normalcy also plays a vital role in driving the Chinese neoliberal machinery with sustained supplies of vulnerable but 'volitional' migrant workers, both sustaining and undermining their ability to thrive and survive in the contemporary Chinese society (Rofel 2016; Litzinger 2016;

Rojas 2016). For millions of dream-pursuing migrants in Shanghai, beyond their awareness and even prior to their start, a set of good-life dreams has prescribed a leave-and-return path for them with almost no outlets. Throughout their own good-life-yearning journey, these migrants travel across varying but more or less scripted routes: as children giving up their education to help with family finances, as mothers/daughters quitting their urban jobs and moving back for tendering their underage and aging beloved family members, or as bread-winners struggling in cities, lonely but hopeful, for supporting their families back at rural homes. Divergent and disparate as their routes might seem to be, the destination remains distressingly identical: regardless of how hardworking they are, the good-life dreams they always turn out to be cruel rather than salvific. Via these hallucinogenic and spellbinding dreams, they are reproduced as volitional subjects willing to work for all but their own well-being—or put more cruelly, as the backbone that shores up the convivial banquet of interests for the Chinese state and capital right upon their premeditated slow death.

Compared with numerous rural migrants in Shanghai, holding a graduate degree from an elite university and the government-issued *Hukou* that officially recognises my 'Shanghaineseness' definitely provides me with much more leverage. Even so, I could not find a job with a good enough salary for even a small condo to call 'home'. As news reports repeat recently, myriads of young people are 'escaping from Bei Shang Guang' (an acronym for Beijing, Shanghai and Guangzhou, the three major metropolises in China) after finding their hard work would never yield the desired outcomes. Understandably, many of them impute their problems to the 'corrupted' communist bureaucracy. Swung by the dominant discourse of neoliberalism alleging that liberalisation, deregulation and democratisation of economics and politics (as Westerners have done it) would finally create a prosperous and equitable society, they submit to the individualistic idea of human rights and personal freedom that is manipulated by neoliberalism to further the self-serving market rule (Harvey 2005). Distressed by the party-state's failed promise of building social equality and justice, they displace their unfulfilled promise of the aspiring fantasy onto the uncharted 'liberal' Western world. Without much hesitation, I submitted my application materials for several US graduate programmes, and half a year later, set on the journey to 'the land of opportunities' with my renewed aspiring fantasy of a good life.

RACIALISED SELF-REFORMATION FOR OTHERWISE

As a new identity category imposed onto my body by the US politics and culture, race became the most significant caliber of my identity and intellectual transformation. When I checked the box that said 'Asian/Pacific Islander' in order to gain entry into US territory, though I had little clue of what this would mean to me, my body was already invested with new biopolitical forces that would later translate into tangible effect and affect to dis/orient me in the 'new' world.

As Howard Winant suggests, race is 'a concept that signifies and symbolises sociopolitical conflicts and interests in reference to different types of human bodies' (2004, 155). Sally Kitch (2009) pushes this argument further and suggests that there is a gendered foundation of racial formation in the United States. As a racial category, Asian is also founded on gender and sexual ideology for its construction in the US context. As for Chinese and Japanese, two pioneering immigrant groups later categorised as 'Asian', the manipulation of the discourses about the sexuality of Chinese and Japanese women first classified them as two different ethnic groups, and then conflated them as one race (Lee 2010). In the contemporary United States, David Eng (2001) posits that the intersection of race, gender and sexuality creates a social consequence for Asian men which he calls 'racial castration', which often comes with a queering effect that frames all Asian men as 'gay', or more literally, sexually receptive 'bottom'. This intersectional relationship, however, produced novel and even 'freeing' experiences for me. I was happy to find that my questioned gender and sexual identities in China were legitimised by my newly acquired racial identity—my favorite flamboyant and 'girly' shirt that might look problematic and suspicious for other racial groups was acceptable for Asian men like me, as a peer graduate student told me. Ironically enough, racially defaulted as 'queer/bottom', my dream of emancipation from the (communist-foisted) gender and sexual subjugation seemed to be realised, at least preliminarily.

Like for gender, class and sexuality, the biopolitical forces invested in race need to be unpacked through a contextualised and historicised lens to explore how the values attributed to 'Asian' have been shifting with the US politico-economic agenda. Different from its derogatory connotation of a promiscuous 'Yellow Peril' that posed a threat to the US civilisation in the late nineteenth century (Reddy 2003), or polluted the racial purity of white Americanness in the early twentieth century (Nuys 2002), the racial meanings of 'Asian' are said to have been whitened, as the deceptive moniker 'Model Minority' suggests. However, this racial euphemism can be easily debunked through the biopolitical optics. To secure and further its advantaged position in the global industrial hierarchy, the United States has developed various policies since the 1970s to attract technological professionals from other countries, particularly Asia, to strengthen and perpetuate its dominance in the 'knowledge economy' (Ong 2006). The success of this segment of new immigrants reassembled by the state through class and gender—usually well-educated middle-class professional men—is then repackaged in racial terms to conceal the biopolitical essence of neoliberal governmentality: as a racial group, 'Asians' are culturally attuned for market success by their 'tradition' of loyalty to authorities and family, embracing of hard work and education, as well as 'natural talents' for subject matters such as science, technology, engineering and mathematics. Building on these alleged virtues, Asians set up a 'role model' for other racial minorities, such as blacks and Latinos, who, in turn, are ironically homogenised as 'lazy, system milking, uninterested in education', thus not deserving success in the society guided solely by the 'jungle rule' of the market. Mari Matsuda (1996) calls the racial myth of the 'Model Minority' an instrument of the distributive, gender and racial injustice in service of neoliberal restructuring.

As a biopolitical technique of neoliberal governance, the 'Model Minority' has taken earth-jolting effects in the 2016 US presidential election. On the one hand, framing minority groups, such as Mexicans and Muslims, as unlawful immigrants coming to the country either for crimes, manipulation of the welfare system or terrorism—thus posing threat to the good life of 'the American people'—and vowing to take drastic measures to keep them out, Trump not only directly speaks to the failed promise of a good life faced by myriad struggling poor/middle-class Americans but also repackages himself as the person who has the 'magic' to materialise the deferred promise and 'make America great again'. On the other, by attacking the 'undocumented immigrants' for realigning himself with the promise of the good life, Trump creates thick and sticky affective forces that draw other 'legal' immigrants/good-life-pursuers towards him. As a result, many first-generation Chinese Americans, a group usually nonchalant to electoral politics, were motivated and rallied around Trump, and took dramatic measures, such as renting airplanes for publicity, for support. But the irony is, actually Asian immigrants well outpace Mexicans in terms of 'undocumented' growth and have seen the fastest growing rate since 2000.[6] Among the roughly 11 million 'undocumented immigrants' in the United States, the ones coming from China amount to around 268,000, and rank no. 4 among all nationalities, only trailing behind Mexico, Salvador and Honduras.[7]

The above critical examination of the racialised neoliberal biopolitics in the United States might seem coherent as an ex post facto thought. It, however, did not take shape as a result of the theoretical and methodological academic training but instead was sparked by my awareness deriving from the contradictory experience between the racial operation of biopolitics and the lived reality. One day I received an email from a student in our PhD programme, asking for the syllabus of the graduate seminar of statistics that I was assigned to teach in the following semester. The unsolicited request was then confirmed by the course schedule posted online, leaving me in stunned bewilderment. Lacking formal training in statistics, I am not qualified for any courses on the subject matter at all, not to mention a graduate seminar. Simultaneously, this exchange brought back a series of incidents that had happened to me before, in which I was told by other students, partially joking and partially serious, that 'as an Asian man, I don't belong in the class about humanities and social sciences'. But this time, the overwhelming fear and shock rendered me unable to shrug off this destabilising message as just another innocuous coincidence.

Tearing apart my habituated way of sensing, thinking and perceiving through intense bodily reactions, the debilitating racialised positioning of my role in the academic world—which I would by no means fit into—carved out a new horizon for me to rethink the inherent relationship between the social status quo and the identity-nexused intersectional biopolitical system. Just as Amhed astutely shows, 'When the arrival of some bodies is noticed, when an arrival is noticeable, it generates disorientation in how things are arranged' (2007, 164). In this case, it is the noticing of the arrival of my racialised body in a prestructured space that it cannot fit into which triggered my disorientation, and then, new orientation to sense and perceive the world. Obviously, the racialised and gendered values invested in my body already prescribed a path for me unknowingly, even before my arrival. To some, this prescribed path might seem empowering and fulfilling as it lays out a route leading to good job opportunities and economic/professional success, definitely different from the one for their predecessors that led to dead-end jobs of mining, railroad construction or laundry and slow death. These variously paved roads, however, might not be as different as they seem to be—be it an IT engineer, a bachelor railroad builder/gold miner, or a laundry shop/fast-food restaurant worker, the differentially built roads along the racial, gender, class and sexual lines actually lead to the same final destination for satisfying the varying needs of 'free market'. Indeed, this final destination, as Foucault argues, is the raison d'être of neoliberal governmentality that seeks to open up more space for the profit making of capital.

In the United States, contrary to the myth of the 'Model Minority', Asians are far from being 'whitened', and so to speak, unable to join the 'superior' club crowded with Caucasians. Seen from a broader view of the intersectional biopolitical system, new Asian immigrants are stuck in a place not necessarily safer/better than that of their less-affluent predecessors. As the widely circulated discourses of the good-life-threatening China (and as the corollary, Chinese) during the 2016 presidential campaign demonstrate, just like the perverse Chinese coolies who were said to have stolen jobs from white workers in the nineteenth and early twentieth centuries, the 'lawful' and well-educated new Chinese immigrants can be easily turned into a new enemy for angry white voters immiserated by the neoliberal structuring. Taking the opposite side to 'unlawful' immigrants of other racialised groups, Asian in general and Chinese in particular, reinforce and perpetuate the racialised biopolitical ground for abrasive political figures like Trump to forward the global neoliberal control. Pitting different groups against each other and blurring their view to capture the broad picture of the intersectional structure, the ever-shifting contour and metrics of the neoliberal governing system render us unable to see through and beyond the chaotic status quo for substantive changes.

CONCLUSION

Building on the reflexive analysis of the author's cross-China and transpacific good-life-pursuing journey of self-formation and reformation, this chapter unravels the intersectional biopolitical mechanism that has been bolstering and subtending global neoliberal restructuring. Building on and extending Foucault's analysis of the apparatus of sexuality, I argue that the intersection of gender, class, race and sexuality constitute a transnational network of power that undergirds and shores up neoliberal governmentality for shaping the population and individual life in line with the marketising and globalising agenda. In the meantime, this intersectional network, as I illustrate, informs and enables an array of aspiring good-life dreams that help reproduce multifarious subjects willing to work for the interests of capital, oftentimes against their own well-being. In revamping and reconstituting its networked gender, racial, sexual and class metrics, this network rejuvenates and reproduces the promises of the aspiring dreams by projecting the unfulfilled ones onto other terrains/objects to re-create and retain the affective forces to human subjects. Always futuristically oriented and other-place-projected, the fantasmatic dreams, therefore, are continually deferred rather than defeated—a crucial mechanism for the continuation and perpetuation of the devastating global neoliberal control. To retrieve our peaceful dwellings, as Amhed puts it clearly and succinctly, 'we need to dismantle what has already been assembled; we need to ask what it is we are against, what it is we are for, knowing full well that this we is not a foundation but what we are working toward' (2017, 2). Offering this nuanced knowledge of the modus operandi of global neoliberalism as the roadmap, I hope to discern a route within our troubled waters that leads to our dreamland.

NOTES

1. President Trump's Inaugural Address, Annotated. www.npr.org, 20 January 2017. http://www.npr.org/2017/01/20/510629447/watch-live-president-trumps-inauguration-ceremony. Accessed 21 August 2017.

2. Full Text of Xi Jinping Keynote at the World Economic Forum, www.cgtn.com, 17 January 2017. https://america.cgtn.com/2017/01/17/full-text-of-xi-jinping-keynote-at-the-world-economic-forum. Accessed 21 August 2017.

3. President Trump's Inaugural Address, Annotated. www.npr.org.

4. Cheng Li, 18th Party Congress' Third Plenum, Another Turning Point in China's Economic Development, the Brookings Institute, 14 November 2013. http://www.brookings.edu/blogs/brookings-now/posts/2013/11/china-communist-party-third-plenum-turning-point-economic-development#. Accessed 21 August 2017.

In 1953, the Chinese government established a registration system called *Hukou*. In this system, people are classified into urban/rural residents based on the place of birth. It was not put into practice until 1958 when Mao's Great Leap Forward policy failed, which left millions of hunger-stricken people in rural areas struggling for survival. Via an array of state policies, such as food rationing, restricted housing, education and medical care only for urbanites, migration from rural areas into cities then became almost impossible during Mao's reign. This system has been retained though modified by the state since China's reform and opening up in 1978.

Like the *Hukou* system, in Mao's envisioning of "a new China," the regulation of women's reproduction should be within the reach of the state as part of the comprehensive social organization and designing. Directed by his optimistic but baseless anticipation of Socialist China's exponentially growing capacity of agricultural and industrial productivity, "throughout most of the 1950s the Chinese government took a decidedly pronatalist position" and did not readjust its planning focus and promote birth control until after the 1960s (White 1994; Greenhalgh and Winckler 2005). After Deng assumed power in 1978, to reshape the population in line with the marketising initiative, the state rearticulated Mao's birth planning into the one-child policy that engages not simply the quantity but also the quality of the population to energise China's neoliberal transition.

5. Trump Assembles America's "Richest Cabinet," www.bbc.com, 1 December 2016. http://www.bbc.com/news/world-us-canada-38168382. Accessed 21 August 2017.

6. Phippen, J. Weston, "Asians Now Outpace Mexicans In Terms of Undocumented Growth," *Atlantic*, 20 August 2015. https://www.theatlantic.com/politics/archive/2015/08/asians-now-outpace-mexicans-in-terms-of-undocumented-growth/432603/.

7. Vivian Yee, Kenan Davis and Jugal K. Patel, "Here's the Reality about Illegal Immigrants in the United States," *New York Times*, 6 March 2017. https://www.nytimes.com/interactive/2017/03/06/us/politics/undocumented-illegal-immigrants.html?_r=0.

BIBLIOGRAPHY

Amhed, Sara. 2007. "A Phenomenology of Whiteness." *Feminist Theory* 8 (2): 149–68.

Amhed, Sara. 2010. "Happy Objects." In *The Affect Theory Reader*, edited by Melissa Gregg and Gregory J. Seigworth, 29–51. Durham, NC: Duke University Press.

Amhed, Sara. 2017. *Living a Feminist Life*. Durham, NC: Duke University Press.
Attwood, Lynne. 1995. "Men, Machine Guns, and the Mafia: Post-Soviet Cinema as a Discourse on Gender." *Women's Studies International Forum* 18 (5/6): 513–21.
Barlow, Tani. E. 1994. "Theorizing Woman: Funü, Guojia, Jiating (Chinese Women, Chinese State, Chinese Family)." In *Scattered Hegemonies*, edited by Caren Kaplan and Inderpal Grewal, 173–96. Minneapolis: University of Minnesota Press.
Berlant, Lauren. 2011. *Cruel Optimism*. Durham, NC: Duke University Press.
Bray, Francesca. 1997. *Technology and Gender: Fabrics of Power in Late Imperial China*. Berkeley: University of California Press.
Chow, Rey. 2002. *The Protestant Ethnic and the Spirit of Capitalism*. New York: Columbia University Press.
Croll, Elisabeth. 1983. *Chinese Women Since Mao*. Armonk, NY: M. E. Sharpe.
Croll, Elisabeth. 1995. *Changing Identities of Chinese Women: Rhetoric, Experience and Self-Perception in 20th Century China*. Hong Kong: Hong Kong University Press.
Eng, David L. 2001. *Racial Castration: Managing Masculinity in Asian America*. Durham, NC: Duke University Press.
Evans, Harriet. 1997. *Women and Sexuality in China*. London: Polity.
Foucault, Michel. 1978. *The History of Sexuality*, vol. 1: *An Introduction*. New York: Vintage Books.
Foucault, Michel. 2003. *Society Must Be Defended: Lectures at the Collège de France, 1975–1976*. London: Picador.
Foucault, Michel. 2008. *The Birth of Biopolitics: Lectures at the Collège de France, 1978–1979*. Basingstoke, UK: Palgrave Macmillan.
Foucault, Michel. 2011. *The Government of Self and Others*. New York: Picador.
Greenhalgh, Susan, and Edwin Winckler. 2005. *Governing China's Population: From Leninist to Neoliberal Biopolitics*. Berkeley: University of California Press.
Harvey, David. 2005. *A Brief History of Neoliberalism*. Oxford: Oxford University Press.
Honig, Emily, and Gail Hershatter. 1988. *Personal Voices: Chinese Women in the 1980's*. Palo Alto, CA: Stanford University Press.
Kimmel, Michael. 1996. *Manhood in America: A Cultural History*. New York: Free Press.
Kitch, Sally. 2009. *The Specter of Sex: Gendered Foundation of Racial Formation in the United States*. Albany: SUNY Press.
Lee, Catherine. 2010. "'Where the Danger Lies': Race, Gender, and Chinese and Japanese in the United States, 1870–1924." *Sociological Forum* 25 (2): 248–71.
Li, Cheng. 1997. *Rediscovering China: Dynamics and Dilemmas of Reform*. Lanham, MD: Rowman & Littlefield.
Litzinger, Ralph A. 2016. "Regimes of Exclusion and Inclusion: Migrant Labor, Education, and Contested Futurities." In *Ghost Protocol: Development and Displacement in Global China*, edited by Carlos Rojas and Ralph A. Litzinger, 191–204. Durham, NC: Duke University Press.
Louie, Kam. 2002. *Theorizing Chinese Masculinity: Society and Gender in China*. Cambridge: Cambridge University Press.

Lu, Sheldon H. 2000. "Soap Opera in China: The Transnational Politics of Visuality, Sexuality, and Masculinity." *Cinema Journal* 40 (1): 25–47.

Matsuda, Mari. 1996. "We Will Not Be Used: Are Asian Americans the Racial Bourgeoisie?" In *Where Is Your Body? And Other Essays on Race, Gender, and the Law*, edited by Mari Matsuda, 149–59. Boston: Beacon Press.

Nuys, Frank Van. 2002. *Americanizing the West: Race, Immigrants, and Citizenship, 1890–1930*. Lawrence: University Press of Kansas.

Ong, Aihwa. 2006. *Neoliberalism as Exception: Mutations in Citizenship and Sovereignty*. Durham, NC: Duke University Press.

Pascoe, C. J. 2007. *Dude, You Are a Fag: Masculinity and Sexuality in High School*. Berkeley: University of California Press.

Povinelli, Elizabeth. 2012. "The Will to Be Otherwise/The Effort of Endurance." *South Atlantic Quarterly* 111 (13): 453–75.

Reddy, Maureen T. 2003. *Traces, Codes, and Clues: Reading Race in Crime Fiction*. New Brunswick, NJ: Rutgers University Press.

Rofel, Lisa. 2007. *Desiring China: Experiments in Neoliberalism, Sexuality, and Public Culture (Perverse Modernities)*. Durham, NC: Duke University Press.

Rofel, Lisa. 2016. "Temporal-Spatial Migration: Workers in Transnational Supply-Chain Factories." In *Ghost Protocol: Development and Displacement in Global China*, edited by Carlos Rojas and Ralph A. Litzinger, 167–90. Durham, NC: Duke University Press.

Rojas, Carlos. 2016. "'I Am Great Leap Liu!' Circuits of Labor, Information, and Identity in Contemporary China." In *Ghost Protocol: Development and Displacement in Global China*, edited by Carlos Rojas and Ralph A. Litzinger, 205–23. Durham, NC: Duke University Press.

Rosen, Stanley. 2003. "Chinese Media and Youth: Attitudes toward Nationalism and Internationalism." In *Chinese Media, Global Context*, edited by Chin-Chuan Lee, 97–118. London: Routledge Curzon.

Sang, Tze-lan Deborah. 2002. "At the Juncture of Censure and Mass Voyeurism: Narratives of Female Homoerotic Desire in Post-Mao China." *GLQ: A Journal of Lesbian and Gay Studies* 8 (4): 523–52.

Sassen, Saskia. 2009. "The Global City Perspective: Theoretical Implications for Shanghai." In *Shanghai Rising: State Power and Local Transformations in a Global Megacity*, edited by Xiangming Chen and Zhenhua Zhou, 3–29. Minneapolis: University of Minnesota Press.

Song, Geng. 2004. *The Fragile Scholar: Power and Masculinity in Chinese Culture*. Hong Kong: Hong Kong University Press.

Stoler, Ann Laura. 1995. *Race and the Education of Desire: Foucault's History of Sexuality and the Colonial Order of Things*. Durham, NC: Duke University Press.

Wang, Yiyan. 2003. "Mr. Butterfly in Defunct Capital: 'Soft' Masculinity and (Mis)engendering in China." In *Asian Masculinities: The Meaning and Practice of Manhood in China and Japan*, edited by Kam Louie and Morris Low, 41–58. London: Routledge.

White, Tyrsene. 1994. "The Origins of China's Birth Planning Policy." In *Engendering China: Women, Culture, and the State*, edited by Christina K. Gilmartin et al., 250–78. Cambridge, MA: Harvard University Press.

Whyte, Martin King. 2010. "The Paradoxes of Rural-Urban Inequality in Contemporary China." In *One Country, Two Societies: Rural-Urban Inequality in Contemporary China*, edited by Martin King Whyte, 1–25. Cambridge, MA: Harvard University Press.

Winant, Howard. 2004. *The New Politics of Race: Globalism, Difference, Justice*. Minneapolis: University of Minnesota Press.

Wood, Michael. 2006. "Kamula Accounts of Rambo and the State of Papua New Guinea." *Oceania* 76 (1): 61–82.

Wu, Fulong, Jiang Xu and Anthony Gar-On Yeh, eds. 2007. *Urban Development in Post-Reform China: State, Market, and Space*. London: Routledge.

Yan, Yunxiang. 2003. *Private Life under Socialism: Love, Intimacy, and Family Change in a Chinese Village 1949–1999*. Palo Alto, CA: Stanford University Press.

Zhang, Charlie Yi. 2014a. "Deconstructing the Hypermasculine National and Transnational Hegemony in Neoliberal China." *Feminist Studies* 40 (1): 13–38.

Zhang, Charlie Yi. 2014b. "Untangling the Intersectional Biopolitics of Neoliberal Globalisation: Asia, Asian and the Asia-Pacific Rim." *Feminist Formations* 26 (3): 167–196.

Zhang, Charlie Yi. 2016. "Queering the National Body of Contemporary China." *Frontiers: A Journal of Women's Studies* 37 (2): 1–26.

Zhang, Everett Yuehong. 2015. *The Impotence Epidemic: Men's Medicine and Sexual Desire in Contemporary China*. Durham, NC: Duke University Press.

Zhang, Li. 2008. "Private Homes, Distinct Lifestyles: Performing a New Middle Class." In *Privatizing China: Socialism from Afar*, edited by Li Zhang and Aihwa Ong, 23–40. Ithaca, NY: Cornell University Press.

Zheng, Tiantian. 2009. *Red Lights: The Lives of Sex Workers in Postsocialist China*. Minneapolis: University of Minnesota Press.

Part III

EMBODIED LIFE: ERASURE, CONTAGION, IMMUNISATION

Chapter 8

On the Government of Bisexual Bodies

Asylum Case Law and the Biopolitics of Bisexual Erasure

Christian Klesse

A strange kind of peace

Peacefully strolling along the sea
breathing the smell of
water.
Wondering about birds of prey
standing still to watch the rings of my stone
throw
wondering about the birds of prey

shores around an island.
Peaceful isle
of green and pleasantry.
Watching the sea moving
away
moving towards me.

Peaceful walks
along the stony beach
of a sea
collecting marvels and sea shells.

Wondering about the birds of prey.

Peacefully sheltering
whom?

On the day
the birds of prey

put their claws in human flesh
Alfred Hitchcock had not imagined
birds of this shape and strength in his wildest
dreams.
They have infra-red.
They have sleepless helpers.
They have a nation at their command.

But
'they' is the name reserved for the
prey,
'us' is the lorry drivers of the nation,
they mount our
trucks
secretely.

Causing damage
causing collapse
to a social security
that no, no not neoliberalism
and austerity
have ravaged but
'them'.

Coast lines prevented defeat for a thousand years
coast lines are our downfall now.
Pleasant cliffs
lovely borders
watching the camera teams watching the lorries' backs
that is the new cliff-hanger
tension drama on newsnight for
'us'.

'Them' who
'We' bestow
our humanitarianism
onto.
'Them' who are
humanity
'us' who are being abused—
disabused of our humanitarian
dreams.

Any sea can drown
if

only it is deep enough
cold enough
and
watched over
by birds of prey
in and out of uniforms.

—Umut Erel

Umut Erel's poem evokes the brutal violence at the heart of the normalised operations of border and immigration control. Landscape descriptions ('island', 'coastlines', 'pleasant cliffs') and cultural references (such as the one to the BBC production *Newsnight*) suggest that the poem comments on the uncanny workings of a machinery of surveillance and policing designed to prevent nonregistered immigration to the United Kingdom by the Border Force and Immigration Enforcement. Whereas 'birds of prey' conjure up the image of helicopters as symbols of state violence, the poem also alludes to the extension of the efforts of keeping unwanted immigrants out of the country into the minds of a nationalised citizenry. Military technology fuses with a hegemonic politics of othering, expulsion and 'inclusion through exclusion' (Agamben 1998), driven by media practices and political ideologies, such as nationalism, patriotism, scarcity, austerity and/or humanitarianism. 'They have infra-red. / They have sleepless helpers. / They have a nation at their command'. The conglomeration of multiple discourses converge to shore up a division between 'us' and 'them' that runs deep and annihilates the ground for innocent by-standing or 'peaceful walks / along the stony beach / of a sea', a division that alone can rationalise the exertion of violence and the indifference to those left dying in their attempts to enter the country. Violence is named in an image that reveals targeted force and attacks the integrity of the human body: 'On the day birds of prey put their claws in human flesh'. Death is not directly mentioned, and not described as an intentional act, but as a calculated possibility—or maybe even an inevitable consequence—of the everyday operation of both the military and the civil machinery of surveillance through the reference to *drowning* at the end of the poem. 'Any sea can drown / if / only it is deep enough / cold enough / and / watched over / by birds of prey / in and out of uniforms'. Foregrounding the murderous dimension of the state's efforts to regulate the population through immigration control while at the same time blurring boundaries between state function and popular sentiment through a focus on the forging of subjectivity, the poem conveys core aspects of a form of power that theorists, drawing upon Michel Foucault (1990; 1997), have called biopolitics. Analyses of the biopolitics of immigration control have frequently deplored race as its primary lens, focusing on identification, biometrics and immigration management (Pugliese 2010;

Vaughan-Williams 2010; Ajana 2012), migrant labour (Topal 2011; Bird in this volume), media framing (Richter in this volume), detention (Farrier 2011; Seitz 2017; Shakshari 2014a) or asylum politics (Estévez 2014). Focusing on a particular kind of migration—that is, refugees who seek asylum because of persecution due to gender and sexual difference and/or dissidence—I show that both race and sexuality may interact and complement each other in biopolitical strategies. My article thus enhances a similar argument made by Jemima Repo on a primarily theoretical level in this volume, by exploring both the life and the death aspects of biopolitical responses to asylum claimants, who cite bisexuality as the cause of their persecution.

Research into asylum case law in many countries (including the United States, Canada, Australia, New Zealand and the United Kingdom) suggests that bisexuals are at serious risk of having their claims dismissed, because their stories and identities are cast as nonplausible or nonconsequential (Berg and Millbank 2009; Rehaag 2008; 2009; Sin 2015). While positive case decisions of gay male and lesbian claimants have been increasing in some jurisdictions, bisexuals are still likely to find their claims on the grounds of persecution because of their sexuality rejected. While the 'discretion requirement'—that is, the expectation that lesbian, gay, bisexual or trans* applicants have to live 'discrete' lives (or, in other words, to 'stay in the closet') to prevent persecution—has been successfully challenged in many jurisdictions, bisexuals are still alleged to being able to 'pass' without hassle, if they only entered heterosexual relations (Spijkerboer 2013). Bisexual claimants often find it impossible to prove their membership in a 'particular social group'. The fluidity bound up with bisexuality and the lack of acceptance for bisexual identities is at odds with the 'immutability' assumption of sexual orientation models (Sin 2015). The common discrimination of bisexuals in asylum law is a direct outflow of what Kenji Yoshino (2000) calls the 'epistemic contract of bisexual erasure'. The hurdles against making bisexual experience intelligible in the field of law and to materialise a right for asylum for bisexual claimants is part of the regulation of the sexuality of migrants' bodies through biopolitical acts of government with all too often necropolitical consequences.

In this article, I draw on both queer and bisexual theory and recent work on biopolitics and necropolitics that have specifically engaged with questions of migration and/or race/ethnicity and racism (see Mbembe 2003; Puar 2007; Farrier 2011; Haritaworn, Kuntsman and Posocco 2014a). I argue that the asylum system operates as a biopolitical/necropolitical apparatus of governance that produces precarious racialised populations who are treated as disposable and stripped of chances for survival through the operation of a rigid deportation regime. LGBTQI people have historically been particularly vulnerable, because international human rights law did not engage in any substantial manner with gender identity and sexuality for a long time (Petchesky 2000; Morgan 2000; O'Flaherty and Fisher 2008; Kollman and Waites 2009). Within these intersecting taxonomies of race and sexuality, bisexual asylum claimants designate a population among queer asylum seekers who face particular challenges when confronting the racist, heterosexist and homonormative asylum and immigration regime. The responses to bisexual asylum claims and their comparatively low chances for success allow us to pinpoint manifold contradictions within international human rights law and expose the hypocrisy of a system that likes to shroud itself with a halo of empathic appeals to solidarity, human rights and social justice.

FROM BIOPOLITICS TO QUEER NECROPOLITICS: GENDER, SEXUALITY AND POSTCOLONIAL ASYLUM

The legal apparatus of the asylum process that enforces applicants to produce a coherent subjectivity that corresponds with a history of consistent exposure to violence and persecution (which is immediately drawn into doubt resulting in the threats of nonadmission, rightlessness, detention or deportation) can be seen as *biopolitical* in the sense applied to the term by Michel Foucault. Foucault did not invent the concept of biopolitics[1] but developed it into a critical tool of exploring novel articulations of power around bodies, human life and populations within the sphere of governance. Foucault's work has inspired a critical debate among a wide range of theorists, including (among others) Giorgio Agamben (1998; 2005), Roberto Esposito (2008; 2011), Michael Hardt and Toni Negri (Hardt and Negri 2001; Negri and Hardt 2005), Achille Mbembe (2003) and Nikolas Rose (2006), that produced biopolitics as 'an interpretive key to analyse how the production of life is articulated with the production of death' (Lemke 2011, xi). In the following, I draw on some of this work to provide a theoretical framework that helps us understand how race and sexuality interact in the subjection of LGBTQ asylum seekers to biopolitical technologies with all too often deadly consequences. Apart from Foucault himself, I draw on the work of Achille Mbembe (2003) and Jemima Repo (this volume). While the production of death is also emphasised within the concept of biopolitics by Agamben and Esposito, Mbembe's discussion of necropolitics brings this aspect to the fore with the highest emphasis. Although the work of both Agamben and Esposito lends itself to theorising the precarious position of asylum seekers (Agamben (1998, 131–35) and even treats the refugee as a paradigmatic figure of bare life within the contemporary period) (see also Richter in this volume), Mbembe most clearly and explicitly highlights the role of colonialism and colonial racism in bringing about the deadly aspects of modern sovereignty.

References to Jemima Repo's work (in this volume) help shore up my claims about the interconnectivity of race and sexuality in the biopolitics and necropolitics in the asylum process.

Foucault elaborates his concept of biopower in the context of his analysis of modern forms of government to grasp the 'the forms of experience and rationality on which power over life was organized' since the eighteenth century (Sellenart 2004, 370). Governmentality, biopower and biopolitics are all parts of a similar larger and overarching project. For Foucault, governmentality designates a new rationality of power that 'has the population as its target, political economy as its major form of knowledge, and apparatuses of security as its essential technical instrument' (Foucault, quoted in Sellenart 2004, 388).

As Foucault explains in his lecture course *Society Must Be Defended*, biopower dovetails with disciplinary power but uses different techniques. 'Unlike discipline, which is addressed to bodies, the new non-disciplinary power is applied not to man-as-body, but to the living man, to man-as-living-being; ultimately, if you like to man-as-species' (Foucault 1997, 242). He also talks of 'biopolitics of the human race' (Foucault 1997, 243), regulating, ordering, measuring, shaping and transforming human life emerges as a core field of governmental tasks (see Foucault 1990; 1997). However, sovereignty in the modern period continues to exert and establish itself as 'the right of life and death'. Yet in contradistinction to prior modes of sovereignty, biopower is bent towards administering life rather than destroying it. Crome (2009, 55) points out: 'The social body has the right to ensure, maintain or develop its life, and where the sovereign, or sovereign state, demands death this is always a demand issued *in terms of, or with the aim of, the furtherance of life*' [my italics]. This notwithstanding, the calculation of death folds into biopower defined as 'a power to foster life or to disallow it to the point of death' (Foucault quoted in Haritaworn et al. 2014b, 6) The reference to drowning at the end of Umut Erel's poem that conjures up the image of refugee boats sinking in the Mediterranean Sea while attempting to seek shelter on the European continent could aptly be theorised this way.

FROM BIOPOLITICS TO NECROPOLITICS

Mbembe (2003) is adamant that modernity sovereignty has always included both biopolitical and necropolitical regimes and tendencies. The division of people into those who are supposed to live and those who are supposed to die is core to the control of populations within the 'biological field' designated by biopolitics. This entails the delineation of the population through biopolitical acts of administration that imply exclusions at the margins.

Foucault refers to the act and rationality of categorising people for this purpose as racism. For Foucault, racism and state power are closely intertwined since racism provides the only way for states to enact biopolitical power as a power to create and take life (or alternately 'to let die'). 'In a normalizing society, race or racism is the precondition that makes killing possible', argues Foucault (1997, 256), or, in a similar vein: 'Once the state functions in the biopower mode, racism alone can justify the murderous function of the state' (1997, 256). Foucault uses the term *racism* in a broad way in the sense of a 'social racism', that provides the rationality for legitimising the banishing and exclusion of groups of people on different grounds (including criminality, madness, sexual deviance, etc.)—resulting either in 'social death' or actual death—with Nazism providing the prime example of the racial state (1997, 259). Foucault limits himself to exploring the implication of Western concepts of governmentality and sovereignty in the extinction of Europe's 'internal Others'. Although he occasionally refers to questions of colonialism in sweeping references (as shown by Hidefumi Nishiyama and Jemima Repo in this volume), he has been rightly criticised for addressing the analysis of the history of colonialism in more detail (Stoler 1995). For Mbembe (2003), slavery and the exploitation in the plantation economies, colonial terror and imperial warfare are the prime examples of biopolitical experimentation. With these histories firmly in view, Mbembe presents a different assessment of the history of European theories of sovereignty and stresses that '*the generalized instrumentalization of human existence and the material destruction of human bodies and populations*' (2003, 14) has to be seen as an integral part of the Enlightenment legacy. In the face of these atrocities, many of which have been enabled by the designation of a status of infrahumanity to specific racialized groups (Gilroy 2000; 2006), the concept of biopower appears to be limited. Mbembe suggests we talk of necropower to allow for a more focused investigation 'under what practical conditions . . . the right to kill, to allow to live, or to expose to death [is] exercised' (2003, 11). As Repo demonstrates in this volume, Foucault thought primarily of racism when he discussed the death aspect of biopolitics, whereas he considered sexuality to be the primary apparatus allowing for the administration of life, spanning from bodies over sexual behaviours to the population as a more abstract category. However, Repo goes on to claim that sexuality, too, can propel the death function of biopolitics—for example, through queer bashing or targeted killing 'as the ultimate punishment for a sexuality that will not discipline and normalise itself' (see Repo in this volume). As the precarious position of LGBTQ applicants in the asylum process and the creation of a condition of virtual nonintelligibility for applicants with fluid, bi- or pansexual attractions and attachments discussed later in this article show, this is not the only way how death relates to the biopolitics of sexuality. Both racism and heteronormativity (as an assemblage of anti-gay, anti-lesbian, anti-bi and anti-trans* practices) have underpinned practices that mark entire groups of the population as being disposable—and thus killable—and thereby has normalised cruelty, often within the allegedly benign paradigms of democracy, civilisation, peace, development and humanitarianism.

VULNERABILITY, DISPOSABILITY AND KILLABILITY

The academic debate about biopolitics and necropolitics has explored multiple aspects of contemporary political culture (Pieper et al. 2011; Clough and Willse 2011; Campbell and Sitze 2013). Differential distribution of vulnerability to the threat of destruction and the lack of a public culture of mourning for the loss of certain—usually racialised—lives follow a calculus that renders certain populations disposable (Butler 2004; 2009; Butler and Athanasiou 2013). Some of this work has explored how these dynamics have affected Muslim diasporic populations in Europe and North America (Puar 2007; Haritaworn, Tauqir and Erdem 2008). Drawing on the pioneering work of Puar (2007), Haritaworn, Kuntsman and Posocco (2014b, 1) deploy the term *queer necropolitics* to theorise 'the intersection of gender, sexuality, violence and precariousness' as it affects queer and trans* bodies in the face of the death worlds of war and imperialism (see also Haritaworn and Snorton 2013; Aizura 2014). This perspective aims to understand 'the ways in which sexual difference is increasingly absorbed into hegemonic apparatuses, in a way that accelerates premature death' (Haritaworn et al. 2014b, 1). Stressing the urgency of racism and the coloniality of power, the critique of queer necropolitics directs attention to 'the everyday death worlds' that span from the explicit 'sites of death making' shaped by war, imperialism, torture and imprisonment to 'the completely normalized violence of the market' (2014b, 2). This work highlights the divisions created by racism, neoliberalism and the wilful compliance of mainstream LGBTQ organisations with homonationalist discourses (Puar 2007).

Some authors have specifically looked at migration. For example, Topal (2011) argues that the subjection of Turkish immigrants to continuous surveillance and the denial of basic rights regarding wages, taxation, accommodation and mobility plus the exposure to racist attacks and deportation amount to state racism that increases the 'death possibilities' for this group. Farrier (2011) applies a biopolitics/necropolitics framework to his analysis of postcolonial regimes of asylum in the United Kingdom and Australia. Following Agamben (1998: 2005), Mbembe (2003), Gilroy (2000; 2006) and Bauman (1995), Farrier (2011) points to the centrality of the camp dispositive—that is, the enormous significance that rigid practices of deportation and the permanence of a network of asylum and deportation camps have for the discursive construction of the figure of the *asylum seeker* as a precarious and disposable subject marked by infrahumanity. 'The camp represents a permanent space dedicated to the impermanence of its inhabitants; a place where the rule of law is defined by its suspension; where those who do not belong are accommodated' (2011, 13) (see also Seitz 2017; Shakshari 2014a). Camps are symptomatic of the politics of exclusion and deterrence that have guided political approaches to asylum in many countries, including the United Kingdom (Stevens 2004; Kundnani 2007).

Shakshari (2014a; 2014b) shows how asylum claims based on gender identity and sexual orientation are overdetermined by a racist discourse that construes the United States and other Western countries as a safe haven for LGBT people while representing Muslim countries (or the 'Third World') as murderous and hostile towards LGBT citizens. In order to present successful and legitimate claims that underscore the urgency of a 'well-founded fear of persecution', asylum seekers are pressurised to present stories that 'demonise' their country of origin (2014a, 100–101). Neoimperial politics create a paradoxical dynamic. Wars are legitimated in the name of protecting LGBTQ rights (Haritaworn et al. 2008), while those who are forced to flee the murderous violence of the wars are caught in the transnational border and camp regime (where their existence is reduced to bare life and where they are frequently vulnerable to further abuse and threats to their lives). For Shakshari (2014a, 95), this illustrates 'the killability of lives that are simultaneously imbued with and stripped of liberal universal rights; lives that are subjected to *the politics of rightful killing*'. The spectre of '*the politics of rightful killing*', couched within networks of biopolitical and necropolitical strategies, fully reveals the complexity and predicament under which asylum claims based on sexual orientation or gender identity are conducted.

The cruelty of locking up hundreds of asylum seekers who had been victims of torture in UK detention centres, successfully challenged in the courts in October 2017, is a case in point. Reportedly, a bisexual asylum seeker from Nigeria who was beaten, flogged and knifed because of his sexuality was among those who advanced the case in the courts (Taylor 2017).

BISEXUALITY, BISEXUAL ERASURE AND THE ASYLUM PROCESS

Research indicates that in many countries asylum claims based on persecution on the basis of bisexuality are much more likely to be turned down than asylum claims not related to sexual minority status in general or claims based on lesbian and gay identification in particular (Berg and Millbank 2009; Rehaag 2008; 2009; Sin 2015). The relative invisibility of bisexuality in refugee law mirrors its lack of visibility in human rights practice and discourse in general (Rehaag 2009). While bisexuality is nominally included in many documents that deal with human rights from the angle of gender and sexuality, bisexuality-related issues are rarely discussed in detail and do not usually inform policy recommendations. There are, of course, well-founded concerns as to whether labels—such as 'homosexual', 'lesbian', 'gay', 'bisexual', or even MSM (men who have sex with men) or WSW (women who have sex with women)—all of which have currency in Western cultural contexts, can be universalised through cross-cultural application. However, Hemmings (2007) points out that authors within transnational queer studies, where such concerns are frequently raised, direct their criticism with particular rigour and vehemence against bisexuality, whereas they may continue to use 'lesbian and gay' or 'homosexual' as a shortcut to refer to same-sex behaviours globally. 'It is as if "lesbian" and "gay" can be understood as strategic uses of identity, but "bisexual" cannot be', Hemmings observes (2007, 22). '[B]isexuality is framed as having an almost unique capacity for reinforcing dominant sexual discourse' (2007, 22). The pronounced hostility towards bisexuality as a concept, category and identity label in some quarters of queer theory has been noted elsewhere (Angelides 2000; Young 1997). This can be seen as an effect of the widespread marginalisation of bisexuality within social theory at large (see Monro, Hines and Osborne 2017; Klesse forthcoming). The lack of consideration of bisexuality in transnational queer studies extends also to the scholarly work that deals with sexuality, gender identity and asylum. For example, Lewis's (2014) analysis of 'queer asylum' and 'queer anti-deportation activism' in the UK, which is concerned with the normative pressures of the asylum process that aims at establishing authenticity through narrative evidence, generically refers to 'gay and lesbian' or 'queer' claims and claimants in most parts of the text. While she talks of 'lesbian and bisexual women' or 'queer women' in order to press a gender-based argument, she only analyses the implications of the problem concerned (i.e., stereotypical representations) with regard to lesbian women's claims. The question how bisexual women are affected by the flaws of the process is not considered (see also Llewellyn 2017). While bisexuality is occasionally nominally evoked, it keeps lingering in the margins of the text. Few texts consider the position of bisexual claims in relation to and comparison with other 'sexual minority claims' (Millbank 2009; Berg and Millbank 2009; Gambetti 2014; Murray 2014; Vogler 2016) and even less specifically address bisexual asylum seekers' experiences with the application process (Rehaag 2008; 2009; 2010; Sin 2015).

THE PROBLEM WITH 'SEXUAL ORIENTATION'

The precarious position of bisexuality within human rights practice and discourse is aggravated by the fact that the notion of 'sexual rights' has been modelled upon the idea of 'sexual orientation' (see Corrêa and Muntarbhorn 2007), which according to Waites cannot capture all people 'who do not relate sexually only to one gender' (2009, 151)—that is, those people who are commonly considered to be 'bisexual'. Waites (2009) believes that the essentialist ideas underpinning sexual orientation discourse militate against the comprehensibility of 'bisexual' identities and ways of life. Sexual orientation is usually perceived to be an immutable trait of an individual's personality (Murphy 1997; Stein 1999; Wilson and Rahman 2005). Sexual minorities have frequently taken recourse to what Yoshino (2000) calls the 'immutability defense'—that is, a pattern of argumentation which suggests that it is unethical to persecute or punish people, who have no choice regarding their sexual attractions and behaviours, because they are allegedly 'born this way'. Bisexual narratives threaten the logic of 'immutability defences' and are at odds with the rigidity of sexual orientation thinking. Some theorists consider multiplicity, fluidity and noncoherence as the sources of bisexuality's unique queer potential for undoing dominant gender and sexuality categories of Western binary thought (Anderlini-D'Onofrio and Alexander 2014). At the same time, the very same features underpin the precarious position of bisexuality in the field of the law and in the asylum process.

The 1951 Geneva Convention Relating to the Status of Refugees established an international obligation to aid refugees who are persecuted on the grounds of race, religion, nationality, membership of a particular social group or political opinion. Article 1 of the 1951 United Nations refugee convention defines as refugee:

> A person who, owing to a well-founded fear of being persecuted for reasons of race, religion, nationality, membership of a particular social group or political opinion, is outside the country of his nationality and is unable or, owing to such fear, is unwilling to avail himself of the protection of that country; or who, not having a nationality and being outside the country of his former habitual residence as a result of such events, is unable or, owing to such fear, is unwilling to return to it. (UN High Commissioner for Refugees, UNHCR, 1951, restated in 2001.) (quoted in Lewis 2014, 960–61)

The convention itself and national refugee laws that followed the convention did not originally contain any references to persecution on the grounds of gender and/or sexuality. However, the UNHCR's (2008) publication of its official guidelines on claims relating to sexual orientation and gender identity has signalled a growing interest in the treatment of lesbian, gay, bisexual, transgender and intersex refugees and asylum seekers (Lewis and Naples 2014). This includes the recognition of sexual orientation as a cause of persecution in Article 10 of the EU Asylum Qualification Directive 2004 (Council of the European Union 2004) and the publication of specific guidelines for the evaluation of asylum claims based on sexual orientation and gender identity in December 2011 in the United States (Presidential Directive 2011; Vogler 2016). While this development certainly ameliorates the situation of refugees seeking protection from persecution on these grounds, it is has partially been driven by the increasing politicisation and instrumentalisation of LGBT human rights for the purpose of cultural and geopolitical hegemony as parts of neoimperial agendas (Shakshari 2014a; 2014b; Aizura 2014; Haritaworn et al. 2014b; Raboin 2017; Llewellyn 2017).

In the absence of specific provisions for claims on the basis of sexual dissidence or minority status, many claimants who flee persecution because of their sexuality advance asylum claims on the grounds of belonging to a 'particular social group'. In a process starting in the early 1990s, many national or transnational jurisdictions finally accepted sexual orientation and gender identity as relevant grounds of persecution and expanded their legal provisions (see Rehaag 2008). For example, the 2004 EU Qualification Directive suggests that 'a particular social group might include a group based on a common characteristic of sexual orientation' (Article 10.1(d), Council Directive 2004/83/EC, Council of the European Union 2004).

DEFINING 'PARTICULAR SOCIAL GROUPS' AND THE TROUBLES OF BISEXUALS

A review of Canadian asylum court cases (1994–2004) shows that applicants who applied for asylum on the grounds of sexual identity by claiming membership in a 'particular social group' were less likely to get granted asylum if they stated bisexuality as the reason for their persecution compared to those who asserted a gay or lesbian orientation (Rehaag 2008). Similar findings have been reported for the United Kingdom, the United States, Australia, and New Zealand (Rehaag 2009; 2010; Berg and Millbank 2009; Sin 2015).

Asylum seekers who advanced claims related to sexual orientation have deployed different rationales which include a focus on one of the following themes: (a) 'political opinion' (around histories of activism or nonconforming sexual and gender behaviours), (b) 'religion' (relating to provisions regarding religious dissidence in the UNHCR Guidelines on Gender-related Persecution) and, most frequently, (c) 'membership in a particular group'. In the following, I highlight the particularly precarious position of bisexuality with regard to the last rationale.

In the 1993 Canadian Supreme Court Decision, *Ward v. Canada*, Justice La Forest argues that there are three different types of social groups: '(1) groups defined by an innate or unchangeable characteristic; (2) groups whose members voluntarily associate for reasons so fundamental to their human dignity that they should not be forced to forsake association; and (3) groups associated by a former voluntary status, unalterable due to its historical permanence' (quoted in Rehaag 2010, 284). Justice La Forest further suggests that '[t]he first category would embrace . . . sexual orientation' (quoted in Rehaag 2010, 284). This statement welcomes sexual minorities within the asylum law because sexual orientation would qualify as an 'innate or unchangeable personal characteristic' (quoted in Rehaag 2010, 284). This passage has subsequently been met with widespread approval in jurisdictions across the world and has entered the UNHCR (2008) Guidelines on Social Group Claims.

'IMMUTABILITY' AND VISIBILITY

Both of the core concepts that underpin the rationality for legitimising asylum claims (i.e., 'sexual orientation' and 'membership in a particular social group') are defined with an emphasis on 'immutability'. However, the fluidity ascribed to bisexual desire and identities is at odds with such essentialist assumptions, in particular if asylum cases are built on claims to 'membership in a particular group'. As a result, 'bisexual refuge claimants are at serious risk of having their cases improperly assessed because of their life experiences are easily misunderstood' (Rehaag 2008, 61).

Berg and Millbank's (2009) research demonstrates that the identity claims of *all* applicants who claim persecution on the ground of sexual orientation are systematically scrutinised and measured against normative ideas based on stereotypical (mis)representations and unrealistic expectations (cf. Dauvergne and Millbank 2003; Shuman and Bohmer 2014; Bennett and Thomas 2013). The authors' close reading of case decisions reveals the particularly destabilising effects that references to bisexuality tend to have on the identity claims of applicants. Even bisexual behaviours of partners or ex-partners (!) have been given as reasons distrusting the credibility of the identities of women seeking asylum on the grounds of a lesbian orientation.

Bisexuality is at odds with the dominant belief in the fixity of sexual identities and the assumption of a neat match between behaviour and identification. Moreover, bisexuals are often not able to prove that 'their' group is sufficiently 'visible' and 'recognisable' for the purpose of legally qualifying as a 'particular social group'. Visibility is a chief criterion for the definition of 'a particular social group' in many asylum jurisdictions, including US refugee law (Spijkerboer 2013, 221). The burden to prove 'visibility' creates a particular predicament for bisexuals as a group for different reasons (some may have been married, there is a lack of a recognisable 'bisexual culture' in most countries, etc.).

A group put at particular disadvantage here are (lesbian, bisexual or queer) women (Lewis 2013; 2014; Llewellyn 2017). This is because the trope of 'visibility' is modelled upon representations of a linear and coherent identity that stretches across the private/public divide. Gay men's experiences are more likely to meet these requirements than women's life trajectories, which are often more constrained and concealed by gender-related expectations of marriage, reproduction, domestic responsibilities, kinship obligations, propriety, and so forth. Moreover, women's queer cultures are often far less publicly organised and less frequently researched and documented, which means that there is less-accurate evidence available that could 'authenticate' women's claims.

The muddled-up and incoherent frameworks of visibility and discretion have further led to an 'excessive focus on the sexuality of individual claimants' (Lewis 2014, 962). For example, the Czech government has been reported to deploy compulsory sexual arousal tests in which asylum applicants were exposed to different kinds of pornography (Akin 2017). Research suggests that the submission of explicit visual documentation by gay male applicants was examined as acceptable evidence in UK court procedures, although the official guidelines at the time stated that documentary evidence of sexual activities is not required to provide proof of sexual orientation (Lewis 2014, 959).[2] Yet the harsh scrutiny of questioning and the frequent reluctance of immigration officials to accept testimonies of witnesses that did not have sex with the applicant themselves steers asylum seekers and their legal teams towards supporting their cases in this way. Again, a legal culture geared towards the (perceived) expectation to display one's sexuality in such a public way causes particularly serious predicaments for women in the asylum process (Lewis 2014).

'Political asylum claims brought by sexual minorities are often assessed according to some measure of public display, exposure, and/or recognition, and these measures are far from consistent' (2014, 944), argue Shuman and Bohmer. Due to the fluidity of bisexuality that often implies nonlinear life narratives and the lack of permanent bisexual spaces and cultural and political institutions bisexual claimants are among the groups that find it difficult to produce the kind of evidence expected by judges in the asylum hearings. In the light of the discussion above, we should assume that this problem is aggravated for bisexual women (compared to bisexual men), but so far no comparative data has been collected (see Rehaag 2009).

Research suggests that many asylum seekers adopt proactive strategies in order to pre-emptively deal with these challenges. This includes 'going public' (e.g., through the usage of media) or 'looking for organisational support' (which may involve forms of voluntary activism which also creates affiliation with and visibility within LGBTQ contexts) (Akin 2017). Many organisations in the field of LGBTQ asylum support operate within an ostensibly narrow framework as indicated by the naming of groups (e.g., the 'UK Lesbian and Gay Immigration Group', cf. UKLGIG 2010, 2013 or the 'Lesbian Immigration Support Group'). The lack of consistent references to and awareness of bisexuality within the wider organisational LGBTQI asylum support spectrum may inhibit bi-identified asylum seekers from looking for organisational affiliation as a means of support in the asylum process. They may also be less likely to be encouraged to build a claim based on a bisexual narrative.

THE 'LGBT' ACRONYM: DISCOURSE AND ADVOCACY

The 'presentation of an internal identity with a static and linear sexual desire is a Western construct, and necessary for the validation of queer asylum narratives in legal terms' argues Deniz Akin (2017, 467). Lesbian and gay identity claims are more easily recognisable and authenticatable within these registers than a bisexual identity marker, but of course 'bisexuality', too, is one optional strategic identity claim. Although being an inherently instable and precarious category, the B is part of a wider, transnationally recognisable 'LGBT trope' within the Western human rights discourse that certainly sits uneasily with individual or collective local identifications. Murray suggests that 'in most Canadian Immigration and Refugee Board and socio-legal discourses, "sexual minorities" tend to be identified or described as homosexual, lesbian, gay, bisexual and/or transgender, such that individuals who make a claim based on persecution as a member of this "particular social group" do so through the utilization of LGBT terminology' (2014, 463).

Compared to previous US governments, the Obama administration took a rather progressive stance on questions on LGBTQ rights and LGBTQ-related humanitarian matters (Lewis and Naples 2014). Some researchers suggest that this policy change resulted in major improvements for LGBTQ asylum claimants, including those groups that have reportedly faced serious difficulties in the asylum process. For example, Vogler (2016) reports that many of the lawyers he interviewed for his project claimed that they had successfully defended bisexual, trans* and gender-queer claimants. He reads this as evidence for a loosening of the rigidity in which identity categories are applied resulting from increased training and awareness-raising within the legal profession and the deployment of new, innovative and assertive legal strategies. In a stark reversal of the queer argument of normativity as constraint, Vogler hails the asylum process as a site for the proliferation of new sexual and gender identities that exceed the Euro-American cultural horizon: '[T]he flexibility of the "particular social group" category allows petitioners to stake out new claims based on their unique sexualities' (2016, 858). Vogler attributes to the asylum system a truly queering potential. Through the recognition of queer identities, it 'increasingly blurs any (imagined) crisp dividing line between heterosexuals and homosexuals, as bisexuals and even those with imputed and ambiguous sexual identities are officially recognized by the state' (Vogler 2016, 885). While I appreciate Vogler's attention to indicators of change, he certainly overstates his case when he claims that 'concerns about defining sexuality as "immutable" rarely affect applicants in practice today' (2016, 882). This contradicts the court experiences of many asylum seekers as evidenced by other researchers (Sin 2015; Llewellyn 2017; Shuman and Hesford 2014). I am not convinced that the existence of new guidelines (e.g., US Citizenship and Immigration Services 2011) and a small number of positive case decisions (however welcome) really signal a watershed. Research in other countries indicates that existing guidelines are often poorly applied and understood (Bennett and Thomas 2013; Singer 2015). Media sources suggest that asylum judges continue to reject bisexual claimants' narratives even in the face of compelling evidence (Stern 2016).

Moreover, the end of the Obama administration and the electoral success of Donald Trump cast serious doubts over Vogler's optimistic predictions. Trump, who counts white supremacists among his advisors, has implemented draconian anti-immigration measures (including executive orders to suspend the State Department's Refugee Assistance Program and implement travel bans for the citizens of several Muslim-majority countries) and has pulled LGBT antidiscrimination measures under the banner of 'religious liberty' and heterosexual family values, is likely to have disastrous implications for LGBTQ asylum seekers (Gessen 2017; Oppenheim 2017; Alter 2017; Moskovitz 2016).

'WELL-FOUNDED FEAR OF PERSECUTION' AND DISCRETION REASONING

Proving that one belongs to 'a particular social group' is not in itself enough for successfully claiming asylum. What is also identified in the process is whether there is a 'well-founded fear of persecution' (Shakshari 2014a). In her discussion of sexual-orientation-based asylum cases in the United States, Llewellyn (2017) observes that only those cases in which claimants show that they experience consistent attacks that cause significant bodily harm are likely to be successful. The possibility to establish a 'well-founded fear of persecution' depends on the creation of a plausible scenario that projects this experience into the future. The production of a linear and coherent identity narrative that highlights visibility, public recognition and 'out-ness' is thereby tied into a normative futurity imagined as a temporality of continuity (Luibhéid 2002; Sin 2015).

Bisexual people's claims to asylum have also been particularly vulnerable to refutation on the grounds of a normative expectation of 'discretion'. Applications by lesbian, gay, bisexual or transgendered people have frequently been turned down because of the reasoning that they could live without the risk of persecution if they remained discrete, did not 'flaunt' their identities—in brief, if they stayed closeted. Bisexuals are assumed to be able to pass easily, if they only decided to 'hide' behind an official heterosexual relationship.

In the meantime, the 'discretion requirement' for LGBTQ asylum applicants has been successfully challenged in many countries. However, bisexual claimants' cases are still routinely dismissed, because they are believed to be able to keep up a pretence and 'pass' as heterosexual without hassle (Spijkerboer 2013). Bisexual claimants face multiple predicaments in relation to the visibility requirements of their sexual disposition. Berg and Millbank argue: 'Claims for group membership on the basis of bisexuality are particularly challenging for advocates who must take care to fully explore the transition or fluidity in the claimant's sexual orientation and identity and yet at the same time contextualise such variability in terms of the claimant's experience of well-founded fear of persecution (2009, 213). 'Living openly' has emerged as a core requirement for getting recognised as a refugee on the grounds of sexual orientation. In 2010, Chief Justice Lorde Rodger argued in the UK Supreme Court decision *HJ (Iran) and HT (Cameroon)* (2010) that only those applicants who 'live openly' as 'practicing homosexuals' could be seen as belonging to a particular social group as defined by the UN refugee convention. Everybody else is thought to have made a 'voluntary choice of discretion' (Lewis 2014, 961). Those who do not share information on their intimate and sexual lives in public circles are suspected of having chosen 'voluntary concealment', which can be cited as a reason for rejecting their claim. Bisexuals who are in other-gender relationships and who are married to one such partner can easily be cast as failing on the account of 'living openly' as understood within the logic of the Supreme Court.

SEXUAL ORIENTATION AND GENDER IDENTITY—A MIXED BLESSING

Bisexual activist circles have started to campaign against the apparent discrimination of bisexual people in the asylum process. For example, in the

United Kingdom, bisexual organisations such as the Bi Index and BiCon have protested against the deportation of Orisha Edwards, a bisexual Caribbean asylum claimant, back to his home country, Jamaica, in 2015 and 2016 (Bisexual Index 2014; Walters 2015; Duffy 2016; Monro 2015). The campaign was finally successful and secured Edwards's right to stay in the United Kingdom. Yet the UK asylum legal system has remained a hostile place for bisexual claimants (Beresford 2016). The UK Bisexuality Report states:

> Asylum seekers who identify as bisexual may experience even greater difficulty than others in proving their sexual identity to adjudicators (something which is frequently demanded). . . . There is a call for the UK Border Agency to provide distinct guidelines for cases of bisexuality, to avoid it being categorised as between straight and gay rather than as its own social group or part of a broader 'persecuted social group' such as LGBT. Obviously it is vital that those whose lives are in danger due to bisexual identities and 'same-and-other gender' attractions are offered the same help as those who are threatened due to lesbian and gay identities or 'same-gender' attractions. (Barker et al. 2012, 22)

These claims are supported by recent research into the UK asylum process (Giametta 2014). Bisexuals have been called the 'invisible asylum seekers' within European asylum jurisdictions (Peyghambazadeh 2016). It is of high and urgent importance for activist groups who aim to support the quests and struggles of LGBTQI asylum seekers to address the specific challenges faced by claimants who do not limit their partner choice to one gender (that is similar to their own). Such claimants are not only likely to run into specific problems with regard to their asylum cases, they may also find a lack of adequate support within existing networks of solidarity.

Yet how can the injustices towards nonmonosexual people in the asylum system be effectively remedied? One approach would consist in building a sustained campaign for the recognition of bisexuality as a 'third' kind of sexual orientation (next to, and equally valid as, heterosexuality and homosexuality), leading to demand for a 'bisexual right' for asylum. Many of the voices that support the quest of bisexual of nonmonosexual asylum seekers are framed in such a way.

Yet there are also problems with and limitations to such an approach. For example, would it do anything to challenge the obsessive search and digging for the 'truth' of asylum seekers' gendered and sexual subjectivities? Critical legal scholars who work within a queer framework have raised their doubts (Seitz 2017; Murray 2014; Akin 2017). Even Sean Rehaag, whose pioneering work has been highly effective in raising awareness for the marginalisation and discrimination faced by bisexual asylum claimants, doubts whether such a strategy would work. Rehaag (2008) notes that many legal scholars and activists have moved away from strategies that rest on citing membership in a social group due to immutable characteristics towards what he calls an identity-based 'fundamental human dignity approach'. Justice La Forest's second definition for 'membership in a particular social group' in *Ward v. Canada* focused on 'reasons so fundamental to their human dignity that they should not be forced to forsake association' (1993, 284). This reasoning, too, was incorporated in a modified way in the UNHCR Guidelines on Social Group Claims (UNHCR 2008). As case law examples in the United States, Australia and New Zealand demonstrate, the jurisdictions of several countries have started to embrace this alternative logic.

The 'fundamental human dignity approach' sidesteps the 'immutability requirement' and does not rely on the notion of sexual orientation. According to Rehaag (2008), relying on the immutability criterion would tie sexual minority refugee protection too closely to trends within scientific sexual orientation research. After all, sexual orientation research has so far been inconclusive and the claims by sociobiological determinists have been contested by social constructivist positions that highlight the malleability of sexual desires and identities (Stein 1999; Rogers 1999; Klesse 2014). We have already seen that bisexuals in particular face difficulties in advancing their claims via sexual orientation narratives. It does not give much hope that Rehaag's (2010) research indicates that bisexuals are treated unfavourably within the 'human dignity paradigm', too. This seems to underscore the validity of an earlier argument by Rehaag (2008), according to which the essentialism of sexual orientation thinking in asylum law runs so deep that applicants are best advised to explore alternatives to 'membership in a particular social group' in their application (such as persecution on the grounds of political opinion and religion, gender or voluntary association).

The salience of identity reasoning within asylum law imposes serious limitations in particular for bisexual claimants. Sexual and gender identity are a mixed blessing, even if they opened up routes for lesbian, gay, bisexual and transgender people to claim asylum. The problem is not only the prevalence of homophobic, biphobic and transphobic attitudes by adjudicators but also the contradictions at the core of the sexual and gender identity models which govern the rationality of case decisions based on sexual orientation (Spijkerboer 2013). Within heteronormative cultures, sexual identity discourses (inclusive of sexual orientation models) are set up in a way that puts sexual minorities in a position where they are always expected to settle for less, face the paradox of being expected to be both discrete and open about their sexuality at the same time. Their identities are excessively scrutinised, which creates exclusive effects at the margins of categories. As an 'intermediate category', bisexuality assumes a particularly precarious position in sexual orientation discourses. It is disposed to disappear, squashed by the poles of the homo/hetero binary or absorbed into the foggy space in between them. Any attempt to prove authentic bisexuality is not only normative but also likely to be futile, due to the notorious doubt attached to representations of bisexuality.[3]

Poststructuralist critical legal scholarship has effectively shown that the very concept of 'rights' is dependent on the concomitant construction of an ideal-typical bearer of rights (Stychin 1995; Beger 2002). Inclusive radical sexual politics need to aim at transforming, modifying or transcending identity and rights models, wherever this may be possible (Lalor 2015). This urgent task is complicated by the implication of sexuality and sexual (human) rights discourses in racialised discourse, postcolonial politics and homonationalist agendas (Puar 2007; Haritaworn et al. 2014a)

CONCLUSION—BIOPOLITICS, NECROPOLITICS AND THE ASYLUM SYSTEM

Asylum claims are part of transnational and national debates on migration and human rights. While the discourse of human rights is, on its surface, committed to a truly universal humanism, it has always remained incorporated in hegemonic world politics that have used race thinking as a wedge to split the universalism it claims to endorse by creating racialised groups of infrahumanity (Gilroy 2000; 2006). This has enabled the creation of extrajuridical spaces—or *states of exception* in the words of Agamben (1998)—that are at the same time confused with juridicity itself. Biopolitics (in the Foucaultian sense of a governmental concern with the population) and necropolitics (in a Mbembian sense of a disenfranchisement due to colonial and postcolonial racism) have guided the continuous infringement and disintegration post–WWII asylum legislation in many countries. Depending on the individual and geopolitical context, the refusal of asylum to individuals or groups of claimants amounts to a decision over life and death. It implies a judgement regarding the disposability of the people concerned. Quests for asylum by LGBTQI people have been ignored for decades until they were successively incorporated into laws and guidelines regarding asylum applications. As a result of these developments, asylum claims on the basis of gender identity and sexuality had to be articulated in an overcondensed field shaped by multiple and conflicting agendas, including human rights advocacy, LGBTQI social movement activism, nationalist anti-immigration and antiblack government politics, and the cultural ideologies of empire (Haritaworn et al. 2014b). At the same time, the governance of gender- and sexuality-related asylum claims has evolved following a genealogy of Western gender and sexuality knowledges steeped in heteronormativity, binary thinking and monosexism that continues to produce (fictionalised) representations of coherent populations based on essentialist qualities (such as men, women, heterosexuals, gay men, lesbians). Bisexuality emerges as an instable signifier for a phenomenon that is assumed to be temporary, elusive or unreal, which renders it particularly difficult for 'bisexual' claimants to prove their belonging to a 'particular social group' that is supposed to be essential and immutable. This results in a heightened precarity of claimants that refer to or allude to bisexuality in their asylum applications. The following quote by Gambetti in a discussion of Butler and Athaniasou's (2013) book *Dispossession* is applicable to this experience: 'Certain bodies and populations are differentially exposed to vulnerability, invisibility, unintelligibility, or abjection by social, political, economic, and sexual mechanisms of power. The conditions of counting as human hinge upon various matrices of exclusion from the three fundamental forms of possession: property, propriety, and priority' (2014, 256).

NOTES

I am grateful for conversations with Chiara Addis that inspired me to write this article and to valuable feedback by the reviewers Benoît Dillet and Hannah Richter. I further would like to thank Umut Erel for allowing me to reproduce her poem 'A strange kind of peace' within this article.

1. See Esposito (2008) for a discussion of organistic, anthropological and naturalistic early twentieth-century social and political theories that deployed the terminology of 'biopolitics'.
2. The current version of the guidelines (Home Office 2016) state that accounts of sexual activity should not be considered (p. 29) and that audio-visual evidence should not be accepted by case workers (p. 30), whereas sections of sexually explicit nature that are part of written evidence should be redacted prior to submission (p. 31).
3. The concept of 'gender identity' has been seen as an inadequate tool for achieving inclusive strategies of rights allocation by some transgender activists, too. Gender identity foregrounds fixity and categorical nonequivalence. References to 'gender expression', however, could have captured blurred or shifting (trans) identifications, too (Currah, Juang and Minter 2006). These arguments show that identity-based rights allocation is riddled with problems and inevitably spirals into exclusivity.

BIBLIOGRAPHY

Agamben, Giorgio. 1998. *Homo Sacer: Sovereign Power and Bare Life*. Palo Alto, CA: Stanford University.

Agamben, Giorgio. 2005. *State of Exception*. Chicago: University of Chicago Press.

Aizura, Aren Z. 2014. "Trans Feminine Value, Racialized Others and the Limits of Necropolitics." In *Queer Necropolitics*, edited by J. Haritaworn, A. Kuntsman and S. Posocco, 129–47. London: Routledge.

Ajana, Btihaj. 2012. "Biometric Citizenship." *Citizenship Studies* 16 (7): 1–20.
Akin, Deniz. 2017. "Queer Asylum Seekers: Translating Sexuality in Norway." *Journal of Ethnic and Migration Studies* 43 (3): 458–74.
Alter, Charlotte. 2017. "LGBT asylum seekers face extra anxiety under president Trump's travel ban." *Time*, 8 February 2017. http://time.com/4661514/lgbt-asylum-seekers-face-extra-anxiety-under-president-trumps-travel-ban/. Accessed 25 October 2017.
Anderlini-D'Onofrio, Serena, and Jonathan Alexander, eds. 2014. *Bisexuality and Queer Theory: Intersections, Connections and Challenges*. London: Routledge.
Angelides, Steven. 2000. *A History of Bisexuality*. Chicago: University of Chicago Press.
Barker, Meg, Christina Richards, Rebecca Jones, Helen Bowes-Catton and Tracey Plowman, with Jen Yockney and Marcus Morgan. 2012. "The Bisexuality Report: Bisexual Inclusion and LGBT Equality and Diversity." *The Open University: Centre for Citizenship, Identities and Governance and Faculty of Health and Social Care*. http://www.bisexualindex.org.uk/uploads/Main/TheBisexualityReport.pdf. Accessed 23 April 2014.
Bauman, Zygmunt. 1995. *Life in Fragments: Essays in Postmodern Morality*. Cambridge, MA: Blackwell.
Beger, Nico J. 2002. "Putting Gender and Sexuality on the Agenda: Queer Theory and Legal Politics." In *An Introduction to Law and Social Theory*, edited by R. Banakar and M. Travers, 173–88. Oxford: Hart.
Bennett, Claire, and Felicity Thomas. 2013. "Seeking Asylum in the UK: Lesbian Perspectives." *Forced Migration Review* 25 (1) (No. 42): 25–28.
Beresford, Meka. 2016. "The Home Office Quizzed a Bisexual Asylum Seeker on LGBT Terminology during His Interview Process." *Pinknews*, 12 November 2016. http://www.pinknews.co.uk/2016/11/12/the-home-office-quizzed-a-bisexual-asylum-seeker-on-lgbt-terminology-during-his-interview-process/. Accessed 25 October 2017.
Berg, Laurie, and Jenni Millbank. 2009. "Constructing the Personal Narratives of Lesbian, Gay and Bisexual Asylum Claimants." *Journal of Refugee Studies* 22 (2): 195–223.
Bisexual Index, The. 2014. "UK Bisexual Community Outraged by Deportation." http://www.bisexualindex.org.uk/index.php/OrashiaEdwards. Accessed 15 February 2017.
Butler, Judith. 2004. *Precarious Life: The Power of Mourning and Violence*. London: Verso.
Butler Judith. 2009. *Frames of War: When Is Life Grievable?* London: Verso.
Butler, Judith, and Athena Athanasiou. 2013. *Dispossession: The Performative in the Political*. Cambridge: Polity Press.
Campbell, Timothy, and Adam Sitze, eds. 2013. *Biopolitics: A Reader*. Durham, NC: Duke University Press.
Clough, Patricia Ticineto, and Craig Willse, eds. 2011. *Beyond Biopolitics: Essays on the Governance of Life and Death*. Durham, NC: Duke University Press.
Corrêa, Sonia Onufer, and Vitit Muntarbhorn. 2007. *The Yogyakarta Principles: Principles on the Application of International Human Rights Law in Relation to Sexual*

Orientation and Gender Identity. http://www.refworld.org/pdfid/48244e602.pdf. Accessed 6 January 2014.

Council of the European Union, The. 2004. "Council Directive 2004/83/EC." http://eur-lex.europa.eu/LexUriServ/LexUriServ.do?uri=CELEX:32004L0083:en:HTML. Accessed 19 April 2017.

Crome, Keith. 2009. "The Nihilistic Affirmation of Life: Biopower and Biopolitics in the Will to Knowledge." *Parrhesia* 6: 46–61.

Currah, Paisley, Richard M. Juang and Shannon P. Minter, eds. 2006. *Transgender Rights*. Minneapolis: University of Minnesota Press.

Dauvergne, Catherine, and Jenni Millbank. 2003. "Burdened by Proof: How the Australian Refugee Review Tribunal Has Failed Lesbian and Gay Asylum Seekers." *Federal Law Review* 31: 299–342.

Duffy, Owen. 2016. "Bisexual Asylum Seeker Wins Home Office Fight to Remain in UK." *Guardian*, 23 January 2016. https://www.theguardian.com/world/2016/jan/23/bisexual-asylum-seeker-orashia-edwards-wins-home-office-uk-jamaica. Accessed 15 February 2017.

Esposito, Roberto. 2008. *Bíos: Biopolitics and Philosophy*. Minneapolis: University of Minnesota Press.

Esposito, Roberto. 2011. *Immunitas: The Protection and Negation of Life*. London: Polity.

Estévez, Ariadna. 2014. "The Politics of Death and Asylum Discourse: Constituting Migration Biopolitics from the Periphery." *Alternatives* 39 (2): 75–89.

Farrier, David. 2011. *Postcolonial Asylum: Seeking Sanctuary before the Law*. Liverpool: Liverpool University Press.

Foucault, Michel. 1990. *The History of Sexuality*, vol. 1. London: Penguin Books.

Foucault, Michel 1997. *Society Must Be Defended: Lectures at the Collége de France 1975–1976*. New York: Picador.

Gambetti, Zeynep. 2014. Dispossession: The Performative in the Political. By Judith Butler and Athena Athanasiou. Cambridge: Polity Press, 2013 (Book review). *Signs: Journal of Women and Culture in Society* 40 (1): 255–58.

Gessen, Masha. 2017. "How Trump Uses "Religious Liberty" to Attack L.G.B.T. Rights." *New Yorker*, 11 October 2017. https://www.newyorker.com/news/news-desk/how-trump-uses-religious-liberty-to-attack-lgbt-rights. Accessed 22 October 2017.

Giametta, Calogero. 2014. *The Sexual Politics of Asylum: Lived Experiences of Sexual Minority Asylum Seekers and Refugees in the UK*. London: London Metropolitan University.

Gilroy, Paul. 2000. *Between Camps: Nations, Cultures and the Allure of Race*. Harmondswoth, UK: Penguin.

Gilroy, Paul. 2006. *After Empire: Melancholia or Convivial Culture*. London: Routledge.

Hardt, Michal, and Antonio Negri. 2001. *Empire*. Cambridge, MA: Harvard University Press.

Haritaworn, Jin, Adi Kunstman and Silvia Posocco, eds. 2014a. *Queer Necropolitics*. London: Routledge.

Haritaworn, Jin, Adi Kunstman and Silvia Posocco. 2014b. "Introduction." In *Queer Necropolitics*, edited by J. Haritaworn, A. Kuntsman and S. Posocco, 1–27. London: Routledge.

Haritaworn, Jin, and C. Riley Snorton. 2013. "Transsexual Necropolitics." In *The Transgender Studies, Reader 2*, edited by A. Z. Aizura and S. Stryker, 66–76. New York: Routledge.

Haritaworn, Jin, Tamsila Tauqir and Esra Erdem. 2008. "Gay Imperialism: The Role of Gender and Sexuality Discourses in the 'War on Terror." In *Out of Place: Silences in Queerness/Raciality*, edited by E. Miyake and A. Kuntsman, 9–33. York: Raw Nerve Books.

Hemmings, Clare. 2007. "What's in a Name?: Bisexuality, Transnational Sexuality Studies and Western Colonial Legacies." *International Journal of Human Rights* 11 (1/2): 13–32.

HJ (Iran) and HT (Cameroon) v Secretary of State for the Home Department. 2010. UKSC 31.

Home Office. 2016. *Asylum Policy Instruction: Sexual Orientation in Asylum Claims* (Version 6.0, 3 August 2016). https://www.gov.uk/government/uploads/system/uploads/attachment_data/file/543882/Sexual-orientation-in-asylum-claims-v6.pdf. Accessed 22 October 2017.

Klesse, Christian. 2014. "Polyamory—Intimate Practice, Identity or Sexual Orientation?" *Sexualities* 17 (1/2): 81–99.

Klesse, Christian. forthcoming. "Dancing on the Waves or Being Washed Away? Representations of Bisexualities in Liquid Modernity." 20th anniversary special issue, *Sexualities.*

Kollman, Kelly, and Matthew Waites. 2009. "'The Global Politics of Lesbian Gay, Bisexual and Transgender Human Rights: An Introduction." *Contemporary Politics* 15 (1): 1–17.

Kundnani, Arun. 2007. *The End of Tolerance: Racism in 21st Century Britain.* London: Pluto.

Lalor, Kay. 2015. "Making Different Differences: Representation and Rights in Sexuality Activism." *Feminist Legal Studies* 23 (1): 7–25.

Lemke, Thomas. 2011. *Biopolitics: An Advanced Introduction.* New York: New York University Press.

Lewis, Rachael A. 2013. "Deportable Subjects: Lesbians and Political Asylum." *Feminist Formations* 25 (2): 174–94.

Lewis, Rachael A. 2014. "'Gay? Prove it': The Politics of Queer Anti-deportation Activism." *Sexualities* 17 (8): 958–75.

Lewis, Rachel A., and Nancy A. Naples. 2014. "Introduction: Queer Migration, Asylum, and Displacement." *Sexualities* 17 (8): 911–18.

Llewellyn, Cheryl. 2017. "Homonationalism and Sexual-orientation-based Asylum Cases in the United States." *Sexualities* 20 (5–6): 682–98.

Luibhéid, Eithne. 2002. *Entry Denied: Controlling Sexuality at the Border.* Minneapolis: University of Minnesota Press.

Mbembe, Achille. 2003. "Necropolitics," translated by Libby Meintjes. *Public Culture* 15 (1): 11–40.

Millbank Jenni. 2009. "'The Ring of Truth': A Case Study of Credibility Assessment in Particular Social Group Refugee Determinations." *International Journal of Refugee Law* 21 (1): 1–33.

Monro, Surya. 2015. *Bisexuality: Identities, Politics, and Theories*. Basingstoke, UK: Palgrave Macmillan.

Monro, Surya, Sally Hines and Antony Osborne. 2017. "Is Bisexuality Invisible? A Review of Sexualities Scholarship, 1970–2015." *Sociological Review*, 1–19. http://journals.sagepub.com/doi/abs/10.1177/0038026117695488. Accessed 19 April 2017.

Morgan, Wayne. 2000. "Queering International Human Rights Law." In *Sexuality in the Legal Arena*, edited by C. Stychin and D. Herman, 208–25. London: Athlone Press.

Moskowitz, Peter. 2016. "Advocates Fear Disaster for LGBTQ Immigrants under Trump." *Vice*, 22 December 2016. https://www.vice.com/en_uk/article/jpmn3k/advocates-fear-disaster-for-lgbtq-immigrants-under-trump. Accessed 22 October 2017.

Murphy, Timothy F. 1997. *Gay Science: The Ethics of Sexual Orientation Research*. New York: Columbia University Press.

Murray, David A. B. 2014. "The (Not So) Straight Story: Queering Migration Narratives of Sexual Orientation and Gendered Identity Refugee Claimants." *Sexualities* 17 (4): 451–71.

Negri, Antonio, and Michael Hardt. 2005. *Multitude*. London: Hamish Hamilton.

O'Flaherty, Michael, and John Fisher. 2008. "Sexual Orientation, Gender Identity and International Human Rights Law: Contextualising the Yogyakarta Principles." *Human Rights Law Review* 8 (2): 207–48.

Oppenheim, Maya. 2017. "Donald Trump to Become First President to Speak at Anti-LGBT Hate Group's Annual Summit." *Independent*, 12 October 2017. http://www.independent.co.uk/news/world/americas/donald-trump-anti-lgbt-address-hate-group-summit-meeting-first-president-us-homphobia-a7997401.html. Accessed 23 October 2017.

Petchesky, Rosalind P. 2000. "Sexual Rights: Inventing a Concept, Mapping an International Practice." In *Framing the Subject: The Politics of Gender, Sexuality and Power*, edited by R. Parker, R. M. Barbosa and P. Aggelton, 81–103. Berkeley: University of California Press.

Peyghambarzadeh, Zeynab. 2016. "Invisible Asylum Seekers." *Pike & Hurricane: A Foreign Affairs Magazine*, 3 June 2016. http://magazine.ufmalmo.se/2016/06/03/invisible-asylum-seekers/. Accessed 19 April 2017.

Pieper, Marianne, Thomas Aztert, Serhat Karakayali and Vassilis Tsianos, eds. 2011. *Biopolitik—in der Debatte*. Wiesbaden, DE: VS Verlag der Sozialwissenschaften.

Presidential Directive. 2011. "Presidential Memorandum—International Initiatives to Advance the Human Rights of Lesbian, Gay, Bisexual, and Transgender Persons." 6 December 2011. https://obamawhitehouse.archives.gov/the-press-office/2011/12/06/presidential-memorandum-international-initiatives-advance-human-rights-l. Accessed 19 April 2017.

Puar, Jasbir. 2007. *Terrorist Assemblages*. Durham, NC: Duke University Press.

Pugliese, John. 2010. *Biometrics: Bodies, Technologies, Biopolitics*. London: Routledge.
Raboin, Thibaut. 2017. "Exhortations of Happiness: Liberalism and Nationalism in the Discourses on LGBTI Asylum Rights in the UK." *Sexualities* 20 (5–6): 663–81.
Rehaag, Sean. 2008. "Patrolling the Borders of Sexual Orientation: Bisexual Refugee Claims in Canada." *McGill Law Journal* 53: 59–102.
Rehaag, Sean. 2009. "Bisexuals Need Not Apply: A Comparative Appraisal of Refugee Law and Policy in Canada, the United States, and Australia." *International Journal of Human Rights* 13: 415–36.
Rehaag, Sean. 2010. "Bisexuals Need Not Apply: A Comparative Appraisal of Refugee Law and Policy in Canada, the United States and Australia." In *The Protection of Sexual Minorities since Stonewall: Progress and Stalemate in Developed and Developing Countries*, edited by Phil C. W. Chan, 281–302. London: Routledge.
Rogers, Lesley. 1999. *Sexing the Brain*. London: Phoenix.
Rose, Nikolas. 2006. *The Politics of Life Itself: Biomedicine, Power, and Subjectivity in the Twenty-First Century*. Princeton, NJ: Princeton University Press.
Seitz, David K. 2017. "Limbo Life in Canada's Waiting Room: Asylum-seeker as Queer Subject." *Environment and Planning D: Society and Space* 35 (3): 438–56.
Sellenart, Michel. 2004. "Course Context." In *Security, Territory, Population: Lectures at the Collége de France 1977–1978* by M. Foucault, 369–402. Basingstoke, UK: Macmillan.
Shakshari, Sima. 2014a. "Killing Me Softly with Your Rights: Queer Death and the Politics of Rightful Killing." In *Queer Necropolitics*, edited by J. Haritaworn, A. Kuntsman and S. Posocco, 93–110. London: Routledge.
Shakshari, Sima. 2014b. "The Queer Time of Death: Temporality, Geopolitics, and Refugee Rights." *Sexualities* 17 (8): 998–1015.
Shuman, Amy, and Carol Bohmer. 2014. "Gender and Cultural Silences in the Political Asylum Process." *Sexualities* 17 (8): 939–57.
Shuman, Amy, and Wendy S. Hesford. 2014. "Getting Out: Political Asylum, Sexual Minorities, and Privileged Visibility." *Sexualities* 17 (8): 1016–34.
Sin, Ray. 2015. "Does Sexual Fluidity Challenge Sexual Binaries? The Case of Bisexual Immigrants from 1967–2012." *Sexualities* 18 (4): 413–37.
Singer, Debora. 2015. "How Do You Prove You Are gay? A Culture of Disbelief Is Traumatising Gay Asylum Seekers." *Guardian*, 24 November 2015. https://www.theguardian.com/commentisfree/2015/nov/24/gay-asylum-seekers-sexuality-home-office. Accessed 15 February 2017.
Spijkerboer, T. 2013. "Sexual Identity, Normativity and Asylum". In *Fleeing Homophobia: Sexual Orientation, Gender Identity and Asylum*, edited by T. Spijkerboer, 217–38. London: Routledge, 217–38.
Stein, Edward. 1999. *The Mismeasure of Desire*. New York: University of Oxford Press.
Stern, Mark Joseph. 2016. "Court Holds Bisexual Asylum-Seeker Isn't Actually Bisexual, Drawing Withering Dissent." *Slate*, 18 August 2016. http://www.slate.com/blogs/outward/2016/08/18/seventh_circuit_rejects_bisexual_asylum_seeker_deportation_relief.html. Accessed 19 April 2017.
Stevens, Dallal. 2004. *UK Asylum Law and Policy: Historical and Contemporary Perspectives*. London: Sweet and Maxwell.

Stoler, Ann L. 1995. *Race and the Education of Desire: Foucault's History of Sexuality and the Colonial Order of Things*. Durham, NC: Duke University Press.
Stychin, Carl F. 1995. *Law's Desire: Sexuality and the Limits of Justice*. New York: Routledge.
Taylor, Diane. 2017. "Torture Victims Were Wrongly Imprisoned in UK, High Court Rules." *Guardian*, 10 October 2017. https://www.theguardian.com/uk-news/2017/oct/10/torture-victims-were-wrongly-imprisoned-in-uk-high-court-rules. Accessed 25 October 2017.
Topal, Cagatay. 2011. "Necropolitical Surveillance: Immigrants from Turkey in Germany." In *Beyond Biopolitics: Essays on the Governance of Life and Death*, edited by P. T. Clough and C. Willse, 238–57. Durham, NC: Duke University Press.
UKLGIG (UK Lesbian and Gay Immigration Group). 2010. "Failing the Grade: Home Office Initial Decisions on Lesbian and Gay Claims for Asylum." c. Accessed 19 April 2017.
UKLGIG (UK Lesbian and Gay Immigration Group). 2013. "Missing the Mark: Decision Making on Lesbian, Gay, Bisexual, Trans and Intersex Asylum Claims." http://www.uklgig.org.uk/wp-content/uploads/2014/02/Missing-the-Mark.pdf. Accessed 19 April 2017.
UNHCR (United Nations High Commissioner for Refugees). 2008. "Guidance Note on Refugee Claims Relating to Sexual Orientation and Gender Identity." http://www.refworld.org/docid/48abd5660.html. Accessed 19 April 2017.
US Citizenship and Immigration Services. 2011. "Guidance for Adjudicating Lesbian, Gay, Bisexual, Transgender, and Intersex Refugee and Asylum Claims." https://www.uscis.gov/sites/default/files/USCIS/Humanitarian/Refugees%20%26%20Asylum/Asylum/Asylum%20Native%20Documents%20and%20Static%20Files/RAIO-Training-March-2012.pdf. Accessed 19 April 2017.
Vaughan-Williams, Nick. 2010. "The UK Border Security Continuum: Virtual Biopolitics and the Simulation of the Sovereign Ban." *Environment and Planning D: Society and Space* 28 (6): 1071–83.
Vogler, Stefan. 2016. "Legally Queer: The Construction of Sexuality in LGBQ Asylum Claims." *Legal & Society Review* 50 (4): 856–89.
Waites, Matthew. 2009. "Critique of 'Sexual Orientation' and 'Gender Identity' in Human Rights Discourse: Global Queer Politics beyond the Yogyakarta Principles." *Contemporary Politics* 15 (1): 137–56.
Walters, Caroline. 2015. "Bisexual Asylum Seekers." (Collective letter). https://bisexualresearch.wordpress.com/2015/05/20/bisexual-asylum-seekers/. Accessed 19 April 2017.
Ward v. Canada. 1993.
Wilson, Glenn, and Qazi Rahman. 2005. *Born Gay: The Psychobiology of Sex Orientation*. London: Peter Owen.
Yoshino, Kenji. 2000. "The Epistemic Contract of Bisexual Erasure." *Stanford Law Review* 52 (2): 353–461.
Young, Stacey. 1997. "Dichotomies and Displacement: Bisexuality in Queer Theory and Politics." In *Playing with Fire*, edited by S. Phelan, 51–76. London: Routledge.

Chapter 9

A Death-bound Subject

The Gravedigger of the Unmarked Mass Graves in Kashmir

Shubranshu Mishra

In *The Notebooks of Malte Laurids Brigge*, Rainer Maria Rilke writes about his first reflections of Paris and his encounters on its terrain, drawing from the mode of being of the underbelly. Strikingly, he portrays the terror of beauty or the terror underneath the beauty, as he withdraws himself from the beautiful landmarks of the city and is drawn to the sights of the vulnerable, the abandoned and the disposable amidst that beauty. In care, he sees annihilation. In life, he sees death, or a borderline between life and death. Through various figures, objects and the scenes of the city he comes in contact with, he writes about its trials and persecutions. The order of the city is immediately juxtaposed with disorder, as though it exists in complete harmony. Rilke (1990, 3) describes a scene where a man has collapsed on the street, submitting to death. A crowd of bystanders and onlookers surrounds him, as though his death was a mere spectre and he was *always-already* meant to be left to die.

What happened thereafter is forced out of sight to be forgotten and remains unknown. There is a pregnant woman struggling to keep her baby, but there is a maternal care centre around and Rilke assures himself about its function for the promotion of birth, so that one can be conditioned to be the pertinent, and that life will take care of her condition. A military hospital that evokes war and violence is in sight, as well as a baby whose bruised forehead seems to be healing even as the child is sleeping, continuing to survive amidst vulnerability. These sights are an indication of a terrain that is contaminated with fear and horror pervading being and becoming. He comes across a building that does not feature on his city map, as though it was not important: a place to live, inexpensive, one that can only exist in the foul smell of a certain neighbourhood, an existence that is marked out and can remain unknown

(Rilke 1990, 3). There is an existential element in Rilke's prose that gets hold of inherent contradictions of life and death not as mere opposites but as existing beside each other. This work of fiction by Rilke is not so different from the lives of those who inhabit the small hamlets of Kashmir. One of the most militarised regions of the world, Kashmir has often been represented through its picturesque landscape and, in so doing, veiling the violence underneath that landscape.[1]

The beauty is striking indeed, but targeted killings and bomb blasts, alongside strikes, curfews and firing on unarmed protestors and stone-pelters, have become the norm in Kashmir. Deaths have resulted from the bullets of armed rebel outfits and the security forces to silence any dissent or opposition from the general public, and there were those who were caught in the crossfire. Since the 1990s, counterinsurgency has meant that every local Kashmiri is to be viewed with suspicion by the security forces. Bose writes, 'In the eyes of the several hundred thousand soldiers and paramilitary troops flooding the Valley, the whole population was suspect—not just disloyal to India but, much worse, in league with the enemy state across the LOC [Line of Control]' (Bose 2003, 112–13). This makes Kashmir a contemporary place of confinement that reduces the person to her or his immediate identity: that is, the origin of the body. The life is disposable and the death is anonymous and the borderline between life and death produces a death-bound subject, a term coined by Abdul R. JanMohamed (2005) in his study of Richard Wright's works, to understand 'how far and how fine does death penetrate into the capillary structures of subjectivity' (JanMohamed 2005, 2). How does one explain this violence? In beauty, there is a continuous shadow of death, terror and horror as underneath the spectacular valleys in Kashmir lie numerous mass graves, unmarked and unidentified. This chapter elucidates the ways in which death and the fear of death—through the sovereign and anonymous site of death where lie buried numerous human beings—penetrate everyday life to produce death-bound subjects. Through the sites of unmarked and unidentified mass graves in Kashmir and the testimony of a gravedigger of one of the unmarked mass graves, I try to understand the gravedigger, Mushtaaq Chaacha—a subject who lives beside the brutality of everyday life—as the figure of a witness.[2]

Drawing from the work of Abdul R JanMohamed (2005) to explore the figure of the gravedigger as the death-bound subject and illustrating the ways in which Mushtaaq Chaacha bears witness, I extend the concept of bearing witness, as conceived by Agamben (1999), from mere extrapolation to an observation of its modifications in a different context where absolute sovereignty is in operation, creating life which can easily be dispensed with. In order to take forward this conversation, I focus on the continuum of impunity and unaccountability in Kashmir to bring to light the structures of militarisation and acquiescence that create the indistinction between a prison and a street. It is also within the realm of the structure of acquiescence that I explore whether bearing witness is possible.[3] Beginning with a discussion of the revelation of the unmarked mass graves in Kashmir and thereafter introducing JanMohamed's concept of a death-bound subject as a witness, I explore the nature of disposability through the works of Cynthia Enloe (2014) and Ranjana Khanna (2009) and its various meanings as to *dispose of*, *being disposable*, and *to dispose*. Through the testimony of Mushtaaq Chaacha, obtained during field research in Kashmir, this chapter draws attention to how counterinsurgency techniques led to certain individuals being *disposed of* and others being used *as disposable* and forced to enter into a death contract with the state. Through a third meaning of disposability, that is, *to dispose*, it locates the renegotiation of the state's death contract by the gravedigger. By preserving objects of the dead, performing religious rites to bury them with dignity and helping families identify their dead, the gravedigger negotiates a new identity of being the caretaker of the martyrs' graveyard. This chapter demonstrates the ways in which the figure of the gravedigger complicates notions of a true witness, in being both the victim and a witness to death, and raises an important question: Is the true witness the one who is buried or the gravedigger, the one who remembers and turns into a caretaker, obtaining a new identity position? The gravedigger and the site of unmarked mass graves form an excellent archive of knowledge about the construction of a subject and the process of becoming a witness. It reveals to us an archaeology of coercion and resistance, as JanMohamed attempts to unpack, through the construction of death-bound subjectivity.

THE REVELATION OF MASS GRAVES IN KASHMIR

There are six hundred thousand troops stationed in Jammu and Kashmir, the highest soldier-civilian ratio in the world, making it the most militarised region in the world (IPTK 2012). According to the estimate of South Asian Terrorism Portal (SATP), 43,960 people have been killed between 1988 and 2016. Further details of the figures suggest that 23,021 of those killed were 'militants'[4], 14,728 were civilians and 6,206 were personnel of the security forces.[5] The government figures suggest that between 1990 and 2011, 13,221 civilians were killed by militants and 3,642 were killed by security forces.[6] Human Rights Watch (HRW) adds that approximately thirty thousand Muslim Kashmiris have escaped to Pakistan to live as refugees, and thousands have crossed the border to seek arms training (HRW 2006, 1). Local human rights organisations believe that, as of 2012, over eight thousand people have disappeared[7], over seventy thousand people have died and as many as six thousand unmarked graves exist in Kashmir as a result of the security-political-juridical impunity in place since 1989 (IPTK 2012). The contested figures and statistics depict only one of the many dimensions of the conflict in Kashmir that refuses any single story to emerge. The cycle of violence has affected the people on all fronts, physical and psychological, and grief and mourning remain an uninterrupted activity for all those who have seen only loss since 1989.

Ever since the armed insurrection began in the Indian-administered Kashmir in 1989, the security forces, particularly the military and paramilitary, have indulged in human rights violations and have functioned with absolute impunity. The Indian government has enacted the Armed Forces Special Powers Act 1990 (AFSPA) and the Jammu and Kashmir Disturbed Areas Act 1992 (JKDAA) as well as other draconian laws like the Terrorist and Disruptive Activities Act (Prevention) Act 1985 (TADA)[8], and the Jammu and Kashmir Public Safety Act 1978 (JKPSA). These laws created a state of emergency-like situation which allowed security forces to impose curfews for several days together, to pick up any person for interrogation and torture them or make them disappear without any accountability.[9] Under the guise of these laws, which provided immunity from prosecution, men were picked up by security forces from their houses or from the streets, taken to detention centres and tortured. Many of these detainees were last seen in custody and never returned. The security forces have either denied picking them up or have often maintained that these persons crossed the Line of Control to seek arms training and return as fighters.

It was the earthquake of 2005 in the Indian- and Pakistan-administered regions of Jammu and Kashmir that made the mass graves visible to the public eye.[10] These areas, close to the Line of Control, have been under heavy security cover and hence difficult to reach for the general public. After the earthquake, civil society organisations and journalists were allowed to assist in and cover the rehabilitation of the earthquake survivors. It is when villagers came in contact with the researchers of these organisations that they reported the networks of unmarked graves alongside schools, mosques, homes and forests. The villagers narrated the involvement of security forces and spoke of being coerced to bury mutilated bodies themselves, most of which, they believed, were the corpses of locals (IPTK 2012). Several cemeteries for the dead remain a site of memory and mourning, a site of loss of hope and humiliation. However, in juxtaposition to these cemeteries lie unmarked and unknown graveyards where numerous ‘others’ lie buried. These sites represent disposability and namelessness of life or death and yet these sites are surrounded by homes, schools, and parks which bring about a sense of normalcy to death, the death of the other.

The Association of Parents of Disappeared Persons (APDP) published a document, *Facts Under Ground*, in April 2008, which suggested that many missing or disappeared persons might have been buried in these graves (APDP 2008). The organisation conducted research in these areas for over two years, between November 2006 and January 2008, and compiled a list of almost one thousand nameless graves in eighteen villages of Baramulla, Boniyar and Uri in North Kashmir. Thereafter, the International People's Tribunal on Human Rights and Justice in Kashmir (IPTK) conducted its research into several unknown mass graves, including the ones put together by the APDP, in Bandipora, Baramulla and Kupwara between April 2008 and November 2009. The report suggested that although the Indian military and paramilitary forces have claimed that the mass graves in Kashmir include bodies of those killed in combat or armed confrontation, in several cases where exhumations and identifications have taken place, these have been found to be instances of extrajudicial deaths.[11] The security forces claim that those found buried were armed rebels and 'foreign militants' who had crossed the Line of Control and were killed 'lawfully' in armed encounters while trying to infiltrate or travelling into Pakistan from Kashmir to seek arms training (IPTK 2009). Research by the IPTK found that most of the bodies in the graves they recorded were of men and large-scale encounter killings resulted in the creation of these secret graveyards. The security forces coerced selected male residents or professional gravediggers to prepare the graves. Although these gravediggers were not remunerated for their services, they decided to bury the corpses in accordance with Islamic religious customs (IPTK 2009).

The State Human Rights Commission (SHRC) conducted an inquiry for three years, after taking *suo moto* cognizance of media reporting and the claims made by the civil society groups in Kashmir. It released an extensive report in 2012, Inquiry Report of Unmarked Graves in North Kashmir, and admitted the presence of 2,730 bodies, out of which 2,156 remain unidentified. These bullet-ridden bodies in unmarked graves were found at thirty-eight sites in North Kashmir in Baramulla, Bandipora, Handwara and Kupwara districts. Of the dead, a few were defaced, five had only skulls remaining, and at least eighteen graves had more than one cadaver inside them. It confirmed what the IPTK had revealed in its study 'Buried Evidence' (2009). According to the report by SHRC, 'The local police, while handing over bodies to the villagers for burial, would tell them that these were of unidentified foreign militants, but 574 of them were later identified as those of locals by their next of kin. Around 17 were even shifted to their native graveyards at different points of time'.

It goes on to suggest that 'there is every possibility that unidentified dead bodies buried in various unmarked graves . . . may contain the victims of enforced disappearances'.[12] This has been the first official acknowledgement that civilians have been killed and buried in clandestine graves. An eleven-member Special Investigation Team (SIT) under the supervision of a Senior

Superintendent of Police (SSP) from the investigative wing of the SHRC conducted the investigation for over three years. The same commission subsequently ordered a further probe, citing the presence of nearly three thousand more graves, with some containing multiple bodies, in the remote districts of Poonch and Rajouri in the Jammu region. The Jammu and Kashmir Home Department maintained that those buried in the graves were militants and if the families wanted DNA tests, they would have to identify both the graveyard and the exact grave where they think their disappeared relative was buried. This has put the onus on the families of the disappeared to identify the graves of their men from these several thousand clandestine sites in order to find their whereabouts.

As the report 'Buried Evidence' (2009) mentions, these sites are 'massified testaments of state power' which 'seek to produce social death and proscribe remembrance. Acknowledgement and articulation of events that precipitate these deaths are forbidden. Internalization of loss and horror is intended to produce fear and isolation. Keeping alive memory, local communities state, is resistance' (IPTK 2009, 18). These mass graves were created and kept secret, under coercion, by men from local villages who constantly had to fear for their lives. It is within that shrunken space of agency that they forged certain practices to keep the memory of the dead intact. I now consider this zone of social death and the making of a death-bound subject by looking into the work of JanMohamed (2005).[13] Once I have sketched out his observations on the space of social death, I explore the technique of the state to make the subject disposable. This subject is first an excess of the population and meant for use in contingencies, and second, when its use is over, the subject can be thrown away, like a disposable item. The subject, however, can make use of a third aspect, which is *to dispose* by articulating her or his condition and the story of those who were *disposed of*. I explore these responses from the testimony of a gravedigger, Mushtaaq Chaacha, in Kashmir.

SOCIAL DEATH: CONTINUUM OF DEATH AND DISPOSABILITY

JanMohamed, in *The Death-Bound-Subject: Richard Wright's Archaeology of Death*, explores the political functions of death and the coercive threat of death, through a close reading of Wright's literary works, from his first book, *Big Boy Leaves Home*, to his last published novel, *The Long Dream*. In both these works, forming the bookends of Wright's oeuvre, the racialised protagonist is coerced to witness horrific deaths to reach 'manhood' and, by being helpless, is made to share the guilt of the 'violent struggles to death' (JanMohamed 2005, 1). JanMohamed tries to understand what happens to the 'life' of that protagonist who matures under a perpetual, however uncertain, fear of death and how that threat unravels during the course of the protagonist's evolution. The protagonist in JanMohamed's work is, as he puts it, the 'black man or woman living in the South between 1900 and the 1950s' (2005, 5). It is the experience of living or existing in a death space, one which is produced by the master or the state as a disciplinary watch-tower. The process of subject formation, under continuous threat of death, makes a death-bound subject. JanMohamed engages with the structure of slavery as a state of living under a conditionally and repetitively commuted death sentence. This state of living in fear of death, as formulated by JanMohamed through a theoretical lens comprising of biopolitics, phenomenology and psychoanalysis, structures a particular subjectivity and produces a death-bound subject. JanMohamed defines a death-bound subject as 'the subject who is formed, from infancy on, by the imminent and ubiquitous threat of death. The death-bound-subject is a deeply aporetic structure to the extent that he is "bound", and hence produced as a subject, by the process of "unbinding"' (JanMohamed 2005, 2).

This process of disengaging with the never-ending probability of death provides an emancipatory possibility for the racialised subject by renegotiating the death contract, one that binds her or him in subjugation. In this regard, JanMohamed understands Wright's work as a kind of 'archaeology of social-death' and his own work as a genealogy of Wright's archaeology (JanMohamed 2005, 38). He writes about Wright's project as one in which, 'by excavating the repressed layers of the slave's consciousness of bondage, he [Wright] transforms them into an acute awareness of autonomy, in his work, slavery becomes the opposite of what it is initially—it becomes a site of freedom' (JanMohamed 2005, 38).

It is in the probability of emancipation of the slave, overpowering her or his subjugation, that I identify the possibility to further develop Agamben's categories of the *Muselmann* and bearing witness. Through the testimony of Mushtaaq Chaacha, we see that although he enters into a death contract, he negotiates with and confronts power, appropriates the death of the other as his own, and becomes a subject who attempts to renegotiate the terms of his condition. This subject encounters death every day and incorporates it as something that he cannot escape. It is through the account of the gravedigger that this chapter makes sense of the response that a death-bound subject formulates to the condition the gravedigger is surrounded with and subjected to, and a kind of dialectic of death that is established when the gravedigger decides to bear witness.

To explore 'the deeper reaches of the death-bound-subject' and the political significance of life and death, JanMohamed introduces a dialectical impact of death, reworking the Hegelian master-slave dialectic, through the concepts of social-death (thesis), actual-death (antithesis) and symbolic-death (synthesis). He posits that for a slave/subject there is a perpetual threat of actual death, which is a central element of the means of production to labour for the master, and therefore a natural consequence of the fear of actual death is to accept social death. Actual death is the biological death of the subject and she is continuously made aware of its eventuality. This threat induces fear, and in order to escape actual death, the death-bound subject lives a social death, a futureless existence of a second-class alienated subject.

Actual death is both a necessary condition for social death and, at the same time, its likely negation as the fear of actual death compels this subject to live a social death. However, risking actual death can be a liberating experience for the subject and it involves taking on death as a 'property of one's own subjectivity' (JanMohamed 2005, 15). The subject confronts this fear to attain a potential for life, regardless of how much worse it may be, and undergoes a symbolic death. JanMohamed defines symbolic death as being 'constituted by the death of the slave's subject-position as a socially dead being and his rebirth in a different subject-position' (2005, 17). It is through acquiring this agency that the death-bound subject tends to find her or his salvation, thereby reconstructing the political economy of death created by the master. In doing this, the subject finds a third space to rearticulate her or his condition, which is neither suicide nor killing the master. JanMohamed writes,

> the possibility of actual-death is the precondition for the slave's social death in that it is the slave's desire to avoid that possibility, that is, his fear of death, that forces him to 'agree' to become a slave, a socially dead being. However, if the slave is willing to die, if he is willing to risk actualising his postponed death, then that actualization will totally negate his social-death or his enslavement. Symbolic-death, in contrast to these other two forms of death, is constituted by the death of the slave's subject-position as a socially dead being and his rebirth in a different subject-position. Symbolic death takes the form of a sublation: it overcomes and preserves that which it supersedes. (JanMohamed 2005, 17)

For a death-bound subject, death is inescapable as her or his entire existence becomes dependent on that perpetual reality and it is through internalisation that a subject 'can gain access to the potentiality of his life only via his willingness to risk the actualization of his death' (JanMohamed 2005, 283). It is important to highlight here that the existence of mass graves in Kashmir points to the possibility of implementing the actual death by the master, the state, which forced the gravedigger Mushtaaq Chaacha, a death-bound subject, to accept his social death as a mere disposable object, by becoming a collaborator with the security forces and burying the dead brought by them. It is this eventuality of a violent death that acts as a technique to create a fear of death for the gravedigger. However, it is within that zone of social death that the subject, the gravedigger, engendered another subversive subjectivity in which he performed religious burials and preserved the objects of the dead. Over time, Mushtaaq Chaacha decided to challenge the never-ending possibility of his death by bearing witness, by recovering the courage and voice to tell his story, and challenge the terms of his disposability. Therefore, the condition of disposability is something that, it appears, is being challenged, or has the possibility to be challenged, even within social death.

It is within this sphere of social death that the state of disposability can be understood. Khanna (2009, 186) provides three meanings of the term disposable: 'the throwaway (object) in production, the available (income) in reproduction, and the sovereign commandment (over life and death and sexual access)'. The first two pertain to excess in order to suggest first, a 'use and throw' object, a disposable dishware of sorts that would be used and then thrown away as waste (in Khanna's words, a 'disposable camera or disposable diaper') and second, an extra income to spend or save, that is, 'disposable income or disposable assets' (Khanna 2009, 184). The disposable product provides utility for a short while and then it is no longer preserved and does not serve any useful purpose. It just remains meaningless, constituting garbage. The latter is preserved for use, for a later stage. Let us focus on the former. Khanna writes,

> With the connotation of 'availability' hanging over that of the 'throwaway', a potential tension is introduced with certain kinds of disposable objects, especially when they happen to be people. It is true, of course, that the throwaway camera is made for our convenience, marketed to be available whenever we happen to need it and produced at a cheaper cost than that of a permanent camera; in this way, the throwaway does promise availability of some sort. (Khanna 2009, 184–85)

Furthermore, Khanna reminds us of one of the earlier usages of the term 'disposable' to suggest that within this state there was an imagination 'of a willed agency, an "ability," if you will, even a capability attached to humankind rather than just to goods or capital' (Khanna 2009, 185). This attached a productive and useful value to the sovereign as being able to do something. A move from 'disposable' to 'dispose' suggests other etymological meanings of the term. Khanna writes,

> Dis-pose, etymologically, suggests a laying down of something, a placing, but it also relates to a disposition suggesting an utterance, a suitable placing and enframing of things and words, rather than a throwing or throwing away. To dis-pose is an exertion of power by the disposer in the utterance of a disposition

made explicit, or, indeed, in a decision to exercise a control over or a discarding of bodies. (Khanna 2009, 185)

For Khanna, this points to the exertion of the sovereign will to make subjects disposable after having used them for a period of time as desired by the sovereign. She explains this third meaning in the context of the use of women in 'any mobile army and as disposable—available for libidinal excess as well as being discardable objects' (Khanna 2009, 186). Khanna's observations are extremely pertinent for this project because they allow for the conceptualisation of disposability in three key aspects. First, as Enloe mentions, in her *Histories of Violence* (2014) video lecture[14], this involves considering disposability in terms of namelessness and having no recognised history—that is, in our project, disposing of people in mass graves.

Second, as Khanna (2009) postulates, there is a temporary use of disposable items: that is, disposable men as gravediggers to be used for labour, to bury the ones killed, for a period of time and when they are no longer effective enough, to be thrown away or left to die. And third, the reason they are discarded is the act of disposition, not of the sovereign's placing of objects or people, but as Khanna evokes through the notion of willed agency, this is when the gravediggers risk their actual death to move out of their social death. It is a way of responding to a condition the sovereign inflicted on them. The three aspects cover what Odysseos calls 'non-pertinent' people who are outside the ambit of sovereign care, making disposability an everyday function of the modern art of government (2015, 1044).[15]

Enloe's reading of disposability—that is, disposing of people in mass graves—sits perfectly well with Agamben and Levi's analysis of a true witness, the *Muselmann*, as the ones who 'touched bottom' and whose death makes complete witnessing impossible. I, however, unpack it to reveal that the ones who have been buried are not the only ones deemed disposable; the ones who buried them were disposable too. Silverman (2014), in the same *Histories of Violence* video lecture series, says that the disposable are not only the ones who have been killed but also those who have been stripped of everything that constitutes their humanity.[16] Departing from Agamben and furthering the concept of bearing witness, I argue that those who bear witness do so to convey the conditions that they lived through and survived or are enduring at present, thereby rejecting the burden of inauthenticity that their story is positioned at the expense of those who have been forced out of sight. They testify through speech or silences, mourning privately and publicly, remembering by memorialising or chronicling their journey from down to the nethermost on the spectrum of bare life, retrospectively or in an ongoing context of atrocity.

It is the relationship of the subject with death, or threat of death, as JanMohamed outlines, which help us to identify the heterogeneities of bearing witness. Agamben regards the *Muselmann* as the complete witness, existing between life and death, who, however, cannot bear witness because of their desubjectivation. Between the survivor and the dead, the left behind and the disappeared, the disposable and the disposed of, is the figure of *Muselmann* easily recognisable? The predicament of bearing witness, as presented by Agamben, through the sole event of the Holocaust renders certain witnesses more authentic than others, thereby creating a gradation of trauma and suffering. A contextual engagement with the act of bearing witness reveals its multiplicities and heterogeneities. Bearing witness is a powerful act that not only invests people to produce their reality for remembrance but also calls them to responsibility. Bearing witness acquires significance for enacting citizenship, or to challenge its absence, and to be interrogative of the state and its practices. It is the production of the truth of unacknowledged violence.

Drawing on Khanna's work, I argue that the second (disposable men as gravediggers) and the third aspect (act of disposition) reveal the gravedigger of the unidentified mass graves, a death-bound subject, as a true witness as well. The gravedigger is thus a true witness through the sovereign's decision that he might be used and then thrown away, and thereafter through his own relation with the dead, by becoming a caretaker of the mass graves and memorialising the dead, the 'non-pertinent', and recovering a voice to bear witness. Enloe (2014) in her video lecture says that disposability is like namelessness, drawing an analogy between pictures of refugees in camps and stones without names at mass graves. Disposability, she says, is not about being nameless in the eyes of those who are related, family or friends, but in the eyes of those who had imagined the ones as disposable. Targets of unlawful killings, considered to be 'non-pertinent', were disposed of in the mass graves, and the death sites are a grim reminder of the function of the modern state as thanatopolitics or necropolitics. All of this is being carried out not necessarily in authoritarian regimes, but in the name of 'progress, democracy, freedom, choice, efficiency and many other admirable terms', as Silverman (2014) outlines in his video lecture. Related to those who were disposed of in the mass graves are the ones who were coerced to bury them and were also seen as disposable, providing labour for a limited period of time and then are left to be abandoned and made nameless. This coercion to enter into a death contract and internalising the space of social death as an inherent condition is a technique of making certain people 'ineligible for personhood', as Cacho (2012) notes about immigrants and racialised groups, and therefore disposable.

There is fear of death as a form of control that looms large and draws the death-bound subject into that contract to be made available for exploitation. As JanMohamed, through Orlando Patterson's work in *Slavery and Social Death*, reminds us, '[i]t is the systematic reliance on the threat and deployment of actual-death . . . that constitutes . . . a fundamental continuity in the process of coercion' (2005, 5). Once the subject enters her or his social death, she or he can then be used for convenience, to bury the dead, at no extra cost, with no remuneration involved, and removing any possibility of remembrance. But what happens when the subject finds a way to subvert the terms of that contract? Some gravediggers of the unmarked mass graves in Kashmir decided to preserve the memory of those they buried within the space of social death, as I show through the testimony of Mushtaaq Chaacha, by storing their clothes, watches and sacred threads in order to rescue their identity. Some numbered those graves to document the corresponding objects found with them. Some buried the men as per Islamic ethos and performing their *Jenaza*.[17]

Is this anti-disposability an attempt to renegotiate the terms of the death contract? Anti-disposability, for Enloe (2014), means recovering the names from the technique of namelessness, but not only so that one can place the identity on a headstone or plaque. Recovering one's name is to recover ideas and one's voice. It is here, I argue, that the gravediggers reject the identity of a gravedigger as ascribed by the security forces and acquire the subject position of a caretaker. The gravediggers not only challenge the power and its coercive exercise to make them collaborators but, within that pact, they subvert the influence of power on them by memorialising the nameless deaths. This is where the third meaning of disposability, that is, *to dis-pose or disposition*, to generate an agency and an ability, and 'take his death upon himself', emerges (Derrida quoted in JanMohamed 2005, 279).

The third meaning of disposability, in the form of disposition of the marginalised, is the zone of the symbolic death of the gravedigger, and 'allows him to redefine the agent who is in control of his life and labour-power; it is now the slave and not the master who sets the terms of the subject as an economic and value structure; hence symbolic-death constitutes the transformation of the subject-commodity into a subject' (JanMohamed 2005, 281). The gravedigger, Mushtaaq Chaacha, confronts his own participation in this pact to completely reject it, risking his actual death, and bears witness by attributing to the clandestine sites the value of martyrs' graveyards. He becomes their caretaker, reconstituting what was earlier stripped away from his subjectivity. He takes the final call on the terms of the death contract and transfers his knowledge about the ones he buried by assisting the family members in identifying them. This witnessing is indeed about the ones he buried.

It, however, is also about his own dialectic relation with death. In the following section, given the theoretical context provided by JanMohamed, Khanna and Enloe, I suggest that a death-bound subjectivity emerges for the subject, the gravedigger, who is a victim, a performer and a witness to death. Based on my conversations with a gravedigger of an unmarked mass grave in Kashmir, I explore whether the gravedigger overcomes the social death, as a result of his disposability, through the way he performed the burials and bore witness, and by eventually speaking out. I unpack whether, in so doing, the gravedigger is renegotiating the death contract. These burials, as he puts it, remind him of his own perpetual certainty, that is death, and in order to escape this eventuality, the gravedigger takes that leap by acquiring the agency of being a caretaker. Through the restructuring of the death contract, various acts of the death-bound subject may well be instances of subversion and can be understood to reveal the politics of death.

MUSHTAAQ CHAACHA: A DEATH-BOUND SUBJECT

Trachimbord,[18] as I call it, is one of the picturesque border villages in Kashmir and at the same time a garrison town with a heavy military presence, given its proximity to the Line of Control. The district is also home to over one thousand graves containing even more bodies buried in them, and over 90 percent remain unmarked. A small village on the side of Trachimbord highway road, located at about twenty kilometres from Trachimbord town and over seventy kilometres from Srinagar, the summer capital of the Jammu and Kashmir, is home to some of the biggest unmarked graves in the Valley. The site gives the impression that these far-flung places were chosen deliberately, as part of the thanatopolitical design of counterterror, making it a tedious task for those living here to have access to the rest of the region. These sites, these stories, were meant to be hidden. As villagers inform me, there are several men in the village with the same name as Mushtaaq Chaacha, but only one who is the caretaker of the 'martyr's graveyard', one of the ways in which people construct their history or challenge the terms of their dehumanisation, reminding us of the death contract.

Mushtaaq Chaacha, nearly seventy-five years of age, was a farmer before he was coerced to bury unidentified bodies for over three years by the security forces in Kashmir, in consonance with the tradition of the master forcing the slave to consent, through the threat of death. As JanMohamed puts it, 'the master forces the slave to choose, at the point of death, between becoming a slave, that is, choosing his social-death, or dying, that is, choosing his actual-death' (2005, 276). Chaacha narrates: 'The men in olive green came during the day, in the dead of the night, in the winters, even when it was snowing' (personal communication, March 2014). He was made to perform this 'ghost-like task', a phrase he uses to express the horrible nature of labour extracted from him. He lifted the dead bodies from the vehicle carrying corpses, carried them to the graveyard and performed the *Jenaza*. Although coerced into this work, he says, 'burying the dead is an act in the service of God' and therefore he surrendered to the master's subjugation (pers. comm., March 2014). 'I went to the mosque each time and offered the men to the Allah, to soil. I did all of that. I had to do all of that', he recounts in a very jittery voice, often searching for the right word and jostling through memory which is at times unblemished and yet fuzzy in the next instant' (pers. comm., March 2014). He continues, 'There were young ones too', and it seemed he could still feel their trace, 'I cleaned their bodies, their beards, to perform the *Jenaza* and put them in the coffin, whatever I could do to perform the funeral, I did it, as a responsibility' (pers. comm., March 2014).

Mushtaaq Chaacha has buried over two hundred bodies on a hillside, very close to his house, now guarded by barbed wire to secure torments and guard its memories. Alongside the graveyard are a school, a park and a hospital, which give a sense of normalcy to the terrain. Children play cricket in the park as if the mystery surrounding the graveyard is unknown to them or as if they are learning to grasp this as normal life. The normalcy is a stark reminder that those who lay buried were the ones who refused to enter into a contract, were considered to be disposable and were disposed of. These practices bind the subjects to the horror of nameless deaths.

In Mushtaaq Chaacha, one finds a figure who was once an unwilling collaborator, as he became party to the wrongs committed by the state, and a defector, by bearing witness. It is difficult to perceive him as a collaborator because he entered into that relation with a deeply 'contradicted will', a phrase JanMohamed uses to explain how the 'master's task is to destroy the slave-as-a-bearer of himself' (2005, 276). Chaacha did it out of fear, he admits, and recounts the first time a dead body was brought by the police, when his clothes and hands became stained with blood. There was a first time and then it became a pattern, an iteration of brutality and horror, and gradually, this tormenting state of being took a toll on his body and his heart. However, he continued with the task 'whether it was 2am or snowing' (pers. comm., March 2014). This is an example of the power to reduce the life of the subject, thought to be disposable, to nakedness. Looking at his wrinkled face as several layers of stories to narrate, it seems he is surviving his own death, as someone not fully dead and existing only to tell a tale, to memorialise. He says, 'it is a responsibility, a commitment' (pers. comm., March 2014).

The overwhelming experience is a haunting one, reminding Mushtaaq Chaacha of the meaning of (non-)existence and its intervention into daily life. Chaacha recounts, 'People sleep in the night, and I cannot. I am with the ones I buried. Whenever I try to sleep, those scenes come to haunt me. These things, that part of life, cannot be forgotten. It is now in my blood. Even when I walk during the day, some or the other instance comes to me to remind me of its horror. I did it for three years, after all' (pers. comm., March 2014). Mushtaaq Chaacha, in many ways, is like Malte. His descriptions are never divorced from death. He encounters the dead and the dying over and over again, as if the dead are still alive and struggling against their anonymity. Although a 'yet to come' death is a constitutive element of the existence of Chaacha, his experience of death is through the death of those he buried. There is negation of the singularity of this experience. It is strongly relational. For him, a 'yet to come' death is a kind of second death and yet he does not evade it, and at times awaits it. He bears witness to his own death through the death of the other. In that awaiting of a 'futural possibility', to use a phrase from *Being and Time* (1962), the contract will continue until the subject decides to renegotiate or revisit the terms of the contract.[19]

The job has left a deep scar on Mushtaaq Chaacha. He has found himself a victim of the death contract as well. He narrates another experience, one that is related to his last burial and perhaps led to him challenging the death pact.

> Once, the Police asked me to dig eight graves and wait until they deliver the bodies. It was very late in the night and we kept waiting. There was no electricity so we sent a message to the police officer asking for details of when they would arrive. He informed us that they had only four bodies and were waiting for the rest to arrive from the location of the encounter. The officer asked us to close the graves and return home. The following morning, a friend in another village invited me for a meal, and by the time I returned, the police officers, with the help of other villagers, had buried eight men in the same graves we had dug the previous night. One of the villagers told me that he had heard that my nephew, who had gone missing, was one of the men buried in those graves. I did not reopen the grave to identify him. I went to Pakistan later to look for my nephew. (pers. comm., March 2014)

Chaacha could no longer continue with the work of burying these nameless men. He does not think he will be able to encounter his nephew's death. After the earthquake of 2005, the volunteers of civil society organisations visited many villages near the LoC and were then told of these large networks of clandestine graves by the villagers. Mushtaaq Chaacha calls it the 'earthquake of life' (pers. comm, March 2014) which led him to challenge his disposability. As Enloe tells us, 'part of not being disposable is to acknowledge your history and no longer see it to be a matter of shame' (2014). It is through confronting his own acts, although as an unwilling collaborator, that Chaacha recovers his ideas and voice about his experience, making it harder for him to be disposed of by the state. As information started to disseminate, with the work of civil society organisations, many people started to visit, looking for the whereabouts of their disappeared men.

He says, 'Many people came looking for their kin. Few identified the bodies, and many did not. Some took the dead bodies to their village and some, erected tombstones over the graves where I had buried them' (pers. comm.,

March 2014). Chaacha has memories of most of the bodies he buried, memories of their physical features and the clothes they were wearing. He preserved the objects found with them and those 'twin brothers', he recounts, 'who looked identical and the lower half of their bodies was non-existent, all accumulated next to their feet' (pers. comm., March 2014). He says that they probably belong to those who are still waiting for them to return. He recounts an incident when a man from Shopian brought an order from the *Tehsildar* (administrative officer) to identify the body of his dead relative. After the identification, he wanted to take the body but Chaacha pleaded with him, as he did with few others, to not do that, as religion does not allow it. Mushtaaq Chaacha thinks the ones who are now buried do not belong to their families any longer but are in the custody of God. In their *Jenaza*, he offered them dignity.

CONCLUSION

In this chapter, I have tried to examine the agency of death in the construction of subjectivity by seeing the gravedigger, Mushtaaq Chaacha, as a death-bound subject who is constituted by the death of those he buried and finds himself profoundly bound to their memory, which is also his own memory. It is by making the decision to bear witness to those nameless deaths, appearing as the existential moment of sublation, that he challenges the death contract with the master. He risks his actual death to come out of the bounds of social death to enter into a zone of symbolic death, a path towards freedom as JanMohamed (2005) outlined. It is this zone of symbolic death, I argue, where he engages in the 'poetic experience of language', as Derrida (2005, 181) suggests about the act of witnessing: the poetic experience points to the incomplete, at times paradoxical, nature of testimony. It presents us with an excellent archive of a death-bound subject. As someone considered disposable, wherein he was forced to labour for the security forces until he was considered worthwhile, Mushtaaq Chaacha was meant to be thrown away. His resistance to disposability led to a disposition in which he recovered a voice to bear witness to his life being reduced to nakedness and transferring his knowledge about the dead to those waiting for their return and seeking justice. While the dead lay buried, their grief screams out through the softening and deeply haunted narrative of the gravedigger. Their collective testimony is not dependent on one another or at the expense of the other. As the caretaker, the gravedigger appropriates their death, not through sharing the guilt but through mourning and performing their last rites, which ties them together, resolving a certain predicament of death.

NOTES

1. The region of Jammu and Kashmir has been a disputed territory between India and Pakistan since 1947. It is divided between India and Pakistan through the Line of Control (LoC), drawn in 1949, which demarcates the region under their control and administration. Pakistan controls the state of Azad Jammu and Kashmir (AJK) and northern areas of Gilgit and Baltistan, India controls Jammu and Kashmir (J&K); and since 1962 Aksai Chin, in the east, is controlled by the People's Republic of China. This chapter pertains to the Jammu and Kashmir being administered by India and the focus is on Kashmir, which is also referred to as 'the Valley' in the chapter.

2. The interviews were conducted between 2013 and 2015 in Kashmir with the gravedigger and his family members, local villagers and activists, researchers and journalists. Names of some of the locations and respondents have been withheld because of reasons of security and confidentiality or on request for anonymity.

3. Agamben (1999) formulates that the testimony of the survivor stands on the one who did not survive and therefore it points to the impossibility of bearing witness. They were the ones who managed to survive, the ones who lived to tell the story through the ones who did not. He regards the *Muselmann* as the complete witness, existing between life and death, who, however, cannot bear witness because of her or his desubjectivation.

4. The term 'militant' is used in official reports for nonstate actors using violent methods in the context of political violence. I use the terms militant, insurgents, rebels and activists (of insurgent organisations) interchangeably depending on the source or the cited text.

5. SATP is a website documenting cases and maintaining a database of terrorism and low intensity warfare in South Asia. This data is till 10 June 2016: http://www.satp.org/satporgtp/countries/india/states/jandk/data_sheets/annual_casualties.htm (retrieved on 15 June 2016).

6. The article by Randeep Singh Nandal, "State Data Refutes Claim of 1 Lakh Killed in Kashmir" in *Times of India*, published on 20 June 2011 can be accessed here: http://timesofindia.indiatimes.com/india/State-data-refutesclaim-of-1-lakh-killed-in-Kashmir/articleshow/8918214.cms (retrieved on 1 June 2016).

7. The term refers to those people who were allegedly detained by the security forces in Kashmir and have gone missing since then.

8. TADA was scrapped in 1995 and was replaced with the equally draconian Prevention of Terrorism Act (POTA). POTA was repealed in 2004 and replaced with the Unlawful Activities Prevention Act 1967 (UAPA). These laws allow the security officials to detain suspects without trial for a considerable period of time from six months to one year.

9. Laws operating in J&K: AFSPA: For any legal action against a soldier, a prior sanction is required from the Home Ministry of the Indian government. The contents of the act can be accessed here: http://mha.nic.in/sites/upload_files/mha/files/ArmForceSpclPwrAct1990-300513.pdf (retrieved 1 May 2016).

10. There has been no common agreement on the factors that define mass graves and therefore several conflicting definitions have emerged over the years. Jesse and

Skinner (2005) posit that this inconsistency of definitions has arisen 'from two quite disparate emphases: one on the number of bodies it takes to constitute a "mass"; and, secondly, the motive behind the creation of the mass grave' (Jesse and Skinner 2005, 56). The number has ranged from two (Mant 1987; Schmitt 2002) to half a dozen (Skinner 1987). Skinner (1987) goes on to suggest that the way those bodies are placed is enough to reflect the motive behind the existence of these death sites and to make its distinction from mass burials in which the bodies are carefully placed. Public tribunal practitioners, anthropologists and sociologists have analysed dimensions of mass graves and extended the definitions to more than the sheer number of bodies, although number remains important. In the case of Kashmir, the IPTK has defined a mass grave as containing 'more than one cadaver of persons who are identified with physical and cultural characteristics used to legitimate their death by the perpetrators' (IPTK 2009, 29).

11. The report by IPTK goes on to suggest that, as an administrative requirement, Indian law requires probing every violent death and identification of corpses. However, in these clandestine graves, this was a procedure that was not followed and the government authorities remain in a state of denial about the significance of these graves.

12. The SHRC report was taken off the public domain and the information in it was widely covered by the media and the copies were only available with the civil society organisations. The media reports can be accessed here: Special Correspondent, "J&K Human Rights Commission's SIT Confirms 2,156 Unidentified Bodies in 'Mass Graves'", *Hindu*, 22 August 2011, http://www.thehindu.com/news/national/jk-human-rights-commissions-sit-confirms-2156-unidentified-bodies-in-mass-graves/article2379921.ece (retrieved on 9 April 2015). "Unmarked Graves: Rights Group Asks SHRC for DNA Test." *Indian Express*, 30 August 2012. http://archive.indianexpress.com/news/unmarked-graves-rights-group-asks-shrc-for-dna-test/995459/0 (retrieved on 9 April 2015).

13. Report by IPTK, 'Buried Evidence', draws on Abdul R. JanMohamed's *The Death-Bound-Subject: Richard Wright's Archaeology of Death*, and Veena Das, Arthur Kleinman, Mamphela Ramphele, and Pamela Reynolds, eds., *Violence and Subjectivity*. Angana P. Chatterji, Parvez Imroz, Gautam Navlakha, Zahir-Ud-Din, Mihir Desai and Khurram Parvez authored the report.

14. The "Disposable Life" series is part of the "Histories of Violence" project, founded by Brad Evans, analysing violence in the twentieth century through "theoretical, aesthetic and empirical dimensions" The "Disposable Life" series can be accessed here: http://www.historiesofviolence.com/#!special-series/c95v (retrieved on 10 April 2016).

15. Disposability is also a running theme of Henry Giroux's work on violence and neoliberalism capitalism and seen as those who are redundant to be abandoned or incarcerated, in the context of the youth and the marginalised communities, particularly in *Stormy Weather: Katrina and the Politics of Disposability* (2007); *Youth in a Suspect Society: Democracy or Disposability* (2009), and, with Brad Evans, *Disposable Futures: The Seduction of Violence in the Age of Spectacle* (2015). For literature related to disposability, refer to James Ferguson's ethnographic work on mine workers

in the Zambian copper belt and their experience of abandonment in the face of economic decline in *Expectations of Modernity: Myths and Meanings of Urban Life on the Zambian Copperbelt* (1999); and Zygmunt Bauman's *Wasted Lives* (2003) in which he looks at migrants, refugees and outcasts as "human waste" or "redundant" as result of modernisation.

16. Max Silverman's lecture for the "Histories of Violence" project Disposable Life can be accessed here: http://www.historiesofviolence.com/#!full-lectures/cq5w (retrieved on 10 April 2016).

17. *Jenaza* is an Islamic tradition of funeral procession and burial.

18. Name borrowed from Jonathan S. Foer's 2002 novel, *Everything Is Illuminated.*

19. As "yet to come" or "futural possibility", I refer to Derrida and Heidegger. Derrida in *Aporias* (1993) deals with aporias of Heidegger's existential formulation of death in *Being and Time* (1962) in which Heidegger rejects any preconceived, biological, psychological or historical meaning of death and argues that death is not a mere end of existence but reveals something about the ontology of being. It is being-toward-death, an authentic mode of relating to death, an anticipation of one's own death, which is the extreme possibility of Dasien. He distinguishes this extreme possibility from perishing and demising. Heidegger suggests, ruling out death as an actuality, Dasein anticipates its death as a futural, a last outstanding possibility and as a projection but one which exists as an imminent reality. He writes, death is the 'possibility of the impossibility of any existence at all' (ibid., 307) It is this relational understanding to death, in anticipation of a futural possibility, a 'not yet', that makes Dasien 'ahead of himself'.

BIBLIOGRAPHY

Agamben, Giorgio. 1999. *Remnants of Auschwitz: The Witness and the Archive.* New York: Zone Books.

APDP (Association of Parents of Disappeared Persons). 2008. "A Fact-Finding Mission on Nameless Graves & Mass Graves in Uri Area." Report. *Srinagar: Coalition of Civil Society.* https://jkccs.files.wordpress.com/2017/05/facts-under-ground-first-report-on-mass-graves-in-kashmir.pdf. Accessed 20 July 2016.

Bose, Sumantra. 2003. *Kashmir: Roots of Conflict, Paths to Peace.* New Delhi: Vistaar Publications.

Cacho, Lisa Marie. 2012. *Social Death: Racialized Rightlessness and the Criminalization of the Unprotected.* New York: New York University Press.

Derrida, Jacques. 1993. *Aporias.* Palo Alto, CA: Stanford University Press.

Derrida, Jacques. 2005. "Poetics and Politics of Witnessing." In *Sovereignties in Question: The Poetics of Paul Celan,* 65–96. New York: Fordham University Press.

Enloe, Cynthia. 2014, February 3. "Disposable Life—Cynthia Enloe." Lecture. https://vimeo.com/85726222. Accessed 20 July 2016.

Heidegger, Martin. 1962. *Being and Time.* New York: Harper & Row.

HRW (Human Rights Watch). 2006. "'Everyone Lives in Fear': Patterns of Impunity in Jammu and Kashmir." New York: Human Rights Watch.

IPTK (International People's Tribunal on Human Rights and Justice in Kashmir). 2009. "Buried Evidence: Unknown, Unmarked, and Mass Graves in Indian-administered Kashmir: A Preliminary Report." *Srinagar: International Peoples Tribunal on Human Rights and Justice in Indian-administered Kashmir.*

IPTK (International People's Tribunal on Human Rights and Justice in Kashmir). 2012. "Alleged Perpetrators: Stories of Impunity in Jammu and Kashmir." *Srinagar: International Peoples Tribunal on Human Rights and Justice in Indian-administered Kashmir and Association of Parents of Disappeared Persons.*

JanMohamed, Abdul R. 2005. *The Death-bound-Subject: Richard Wrights Archaeology of Death*. Durham, NC: Duke University Press.

Jessee, Erin, and Mark Skinner. 2005. "A Typology of Mass Graves and Mass-grave Related Sites." *Forensic Science International* 152 (1): 55–59.

Khanna, Ranjana. 2009. "Disposability." *Differences* 20 (1): 181–98.

Mant, A. K. 1987. "Knowledge Acquired from Post-war Exhumations." In *Death, Decay and Reconstruction*, edited by A. Boddington, A. N. Garland, and R. C. Jenaway, 65–78. Manchester: Manchester University Press.

Odysseos, Louiza. 2015. "The Question Concerning Human Rights and Human Rightlessness: Disposability and Struggle in the Bhopal Gas Disaster." *Third World Quarterly* 36 (6): 1041–59.

Rilke, Rainer Maria. 1990. *The Notebooks of Malte Laurids Brigge*, translated by Stephen Mitchell. New York: Vintage Books.

Schmitt, S. 2002. "Mass Graves and the Collection of Forensic Evidence: Genocide, War Crimes and Crimes against Humanity." In *Advances in Forensic Taphonomy: Method, Theory and Archaeological Perspectives*, edited by W. D. Haglund and M. H. Sorg, 277–292. Boca Raton, FL: CRC Press.

Silverman, Max. 2014, March 11. "Disposable Life." Lecture. https://vimeo.com/88747506. Accessed 20 July 2016.

Skinner, M. 1987. "Planning the Archeological Recovery for Evidence from Recent Mass Graves." *Forensic Science International* 34 (4): 267–287.

Chapter 10

Biopolicing the Crisis

Gendered and Racialised 'Health Threats' and Neoliberal Governmentality in Greece and Beyond

Dimitra Kotouza

Looking at widely broadcast episodes of sexually and racially targeted policing in Greece between 2012 and 2013, I analyse the biopolitical dimensions of such governmental action in a context of deep political and economic crisis. Since 2009, the Greek state has struggled to manage this crisis and the imposition of a neoliberal restructuring under conditions of lost sovereignty over economic and monetary policy. How are we to understand the concurrency between crisis (which involved the production of extreme levels of unemployment, as well as strong movements of resistance) and the medicalised policing and quarantining of migrant populations and 'dangerous' nonmale sexuality? I argue that the neoliberal management of crisis under a security *dispositif* necessitated a logic of legitimation which reinforced the hierarchical demarcation between the Greek heterosexual male citizen and his gendered and racialised others.

To define concretely the specific types of bodies targeted, as well as the political and affective function of this targeting, I analyse this biopolitical governmental strategy by introducing historical, economic, cultural and psychosocial elements that qualify Foucault's understanding of racial biopolitics, Agamben's sovereign exception and Esposito's concept of immunisation. While all three theorists have emphasised the continuity between totalitarian or Nazi thanatopolitics and modern liberal biopolitics, they have neglected important economic, ideological and psychic dimensions of biopolitics that come into play in the crisis of the neoliberal paradigm. Neoliberal ideology presents the autonomous functioning of the market as a mechanism that immunises the national community. It constitutes marginalised bodies as the justified outcome of market mechanisms and as a threat to the symbolic and physical integrity of the civic body. This was a crisis management technique that facilitated the legitimacy of market-led governmentality through the spectacle of abjection of those marginalised bodies and the policing of borders and sexuality. Yet this technique only partly legitimised the neoliberal restructuring. Its other outcome was to strengthen the social demand for the complete immunisation of the national civic body, which neoliberal governance is structurally unable to guarantee. This dynamic is characteristic of the contemporary shift towards authoritarian politics worldwide.

EPISODES OF GOVERNMENTAL ACTION IN THE CRISIS

In April 2012, Greece was facing not only an economic crisis but also a deep political crisis. Since 2010, there had been successive resignations of government MPs, including that of the then prime minister, George Papandreou, in November 2011. Fierce public and political resistance to the severe austerity contained in the Memorandum of Understanding, upon which depended the International Monetary Fund/European Community/European Union (IMF/EC/EU) bailout and thus the Greek state's solvency, had led to the formation of a caretaker technocratic government under Lucas Papadimos, supported by the Panhellenic Socialist Movement (PASOK), New Democracy (ND) and the Popular Orthodox Rally (LAOS). This temporary government passed a package of severe austerity measures amid mass riots in the centre of Athens,[1] secured a second bailout, and announced elections for May 2012. In that tense conjuncture, the incumbent parties' pre-election strategy focused not on the biggest issue concerning the country, the economic crisis, but on immigration and health.

Opening new detention camps to fit more undocumented immigrants was presented as an issue of extreme urgency, and several army barracks were converted to detention centres in anticipation of more arrests (*To Vīma* 2012b). Meanwhile, opposition parties criticised the disarray in the public health service after a 40 percent cut in the health budget (Kentikelenis et al. 2011), including shortages of cancer medications and reports of an increase in HIV infections. In response to this, the health minister, Andreas Loverdos, launched a public health campaign that explicitly targeted immigrants and sex workers. The campaign was led by the police, enabled by a ministerial directive for the compulsory testing, treatment and potential quarantining of populations deemed 'at risk' of a variety of diseases, including hepatitis, influenza, malaria, polio, syphilis, tuberculosis and HIV, as well as the evacuation of overcrowded accommodation (*eKathimerini* 2012).[2] The police campaign was launched sensationally as the fight against a 'public sanitation bomb' that 'threatened' Greek citizens (*To Vīma* 2012a). To contain this 'threat' in accordance with the new directive, the public order minister ordered the arrest and forced medical examination of hundreds of immigrants who lived in overcrowded flats, and the evacuation of their homes. Undocumented immigrants were taken to detention centres.

As part of the same campaign, over a hundred female drug users suspected of sex work were also arrested and subjected to HIV rapid tests without consultation or consent. Twenty-seven of those who were found to be sero-positive were charged by the state prosecutor with intentional grievous bodily harm against their clients—'family men' as Loverdos, and, by extension, the media called the 'victims' of this supposed crime (Fyntanídou 2012).[3] The women's photographs, personal details, including often the names of their relatives and their workplace addresses, were then published online by the police, and were reproduced on newspaper covers and prime-time television news, calling on 'family men' to get tested for HIV if they recognised any of the women. A Russian woman's details were exposed first. This was reported as a vindication of Loverdos's assertion that HIV was transmitted 'from the illegal immigrant to the Greek client to the Greek family' and his call to 'deport' sex workers (Chatzīgeōrgiou and Vergou 2011). Yet the great majority of women with the virus turned out to be Greek. They were detained for eight months without medical treatment and were released when the charge of grievous bodily harm could not be substantiated (Eleutherotypía 2013a).

The elections brought the first breakthroughs of the Coalition of the Radical Left (SYRIZA) and the neo-Nazi party Golden Dawn (GD), but a new coalition government including PASOK, ND and DIMAR (Democratic Left, a small party that had split off from the SYRIZA coalition) was able to be formed in June 2012. As GD was gaining favourable exposure in nationwide television due to its sudden electoral rise, the newly appointed minister of public order, Nikolaos Dendias, launched an operation against undocumented immigrants provocatively named 'Xenios Zeus',[4] with the stated aim of 'reducing crime' in Greek cities. Mass stops and arrests of nonwhite persons were carried out to certify their documents, which, on several occasions, involved unlawful detentions and abuse not only of immigrants but also of nonwhite tourists. The vigilante participation of GD proactively assisted the police by delivering those 'suspects' to police stations. According to Human Rights Watch (2013), almost eighty-five thousand individuals were apprehended under the Xenios Zeus operation between August 2012 and February 2013, but fewer than 6 percent were found to be in Greece unlawfully.

Meanwhile, the number of homophobic and transphobic street assaults increased greatly, along with arbitrary arrests and police abuse, according to LGBTQ organisations (Eleutherotypía 2013b; Galanoú 2013). The public order minister responded to questions regarding the indiscriminate arrests of trans women by defending his campaign to 'enhance the citizens' sense of safety and to improve the general image of the areas' (Dendias 2013, 1) as well as to stop sex trafficking, associating trans women with the sex trade (ibid., 2). Interestingly, after the arrests of GD leaders in September 2013 for the murder of the Greek antifascist Pavlos Fyssas, the number of assaults against immigrants fell, while the number of homophobic and transphobic assaults continued to rise (Fōtiadī 2014; DOL Newsroom 2014).

All of these developments led the majority of centrist and left commentators to speak of a political shift to the far right. Yet, beyond naming, we also ought to ask why, in this conjuncture of economic and political crisis, the government prioritised the policing of borders, health and sexuality, and whether this represents the arbitrary exercise of political power or a specific form of governmental rationality that responded to the conjuncture of crisis.

CRISIS, POLICING AND HEGEMONY

The title of this chapter, alluding to *Policing the Crisis* by Stuart Hall et al. (1978), would recommend analysing the political strategy of these governments by paying special attention to their ideological and social control apparatuses. As with the moral panics over 'mugging' in 1970s Britain, in this case, panic was cultivated over 'illegal' immigrants, who came to stand in for all 'crime', as well as over anyone thought to endanger the traditional patriarchal heterosexual family. Following the Gramscian paradigm of Hall et al., these moral panics chimed with the dominant conservative common sense in Greece at a moment of hegemonic crisis. The mass, heterogeneous anti-austerity Movement of the Squares in the summer of 2011 had led to the collapse of the previous government of PASOK, and the large confrontational demonstrations of February 2012 had led to the call for the May 2012 elections. The emergence of the law-and-order state (ibid., 273–326) and its propaganda machine can thus be understood as a response to this political crisis.

Yet, unlike 1970s Britain, the historical moment of this crisis has not been one of intensified class struggle between an organised working class and a state that seeks to break its power. On the contrary, the movements of this period were interclass and defended national conceptions of democratic citizenship (Kotouza 2011). Despite the intense police repression of these movements, the policing campaigns described here targeted populations that did not and could not take part in the movements against austerity because of their already existing separation from Greek 'mainstream society'. Already marginalised and unable to form cohesive communities, they had little capacity to develop political self-organisation, in contrast to black and Asian communities in 1970s Britain. These were thus not populations that posed a political threat.

Further, while the concept of 'moral panic' shows how the discourses around crime and prostitution served to gain back a hegemonic discursive position for the government, the questions remain of (a) whether this is merely the arbitrary exercise of political power over othered populations, or if, on the contrary, it represents a specified governmental rationality that responds to the problems of public health and crime; and (b) why specifically the border and sexuality have been the major points of governmental concern with regard to the population in this moment of crisis.

SECURITY, BIOPOLICING AND NEOLIBERAL GOVERNMENTALITY

To put these questions in the terms of biopolitics, we might ask to what extent these governmental actions are rational attempts to 'make live'—that is, to make the population healthy and productive by preventing disease and crime—or if the neoliberal government of crisis entails a different motivation. In the context of mass unemployment affecting one-fourth of the population and a broader decline in health-care provision, how can these measures be understood as the performance of the governmental function of 'making live', when in fact the economic measures pursued diminished this population's means of life? Unemployment and the decline in public health care were the direct results of the neoliberal economic policies imposed on Greece by the IMF and the European Commission. The governments of this period, while at times attempting to negotiate less severe cuts, eventually imposed them using the full force of the police. What was then the role of the policing of 'others' in this context?

To answer this question, it helps to clarify what the meaning of 'population' is in the analysis of biopolitical power. Robert Mitchell (2016) notes that in Foucault's work there is an ambiguity regarding whether 'population' has primarily a biological or a political meaning. Foucault (2003) has defined the population upon which biopolitical power is applied as 'man-as-species' (242)—a biological 'multiple body' (245) irrespective of national borders. Yet, in his analysis of new security mechanisms and governmental thought in the eighteenth century (Foucault 2009), it becomes clear that this body must also refer to the biological aspects of the population within a specified national territory (see esp. 74–75). Foucault's emphasis on the universality of 'man-as-species' might imply his anticipation that the biopolitical management of population on a world scale is the dominant tendency of biopolitical rationality. Still, the tension between the regulation of health and productivity of the species versus that of a specified national population defined by ancestry or birth within a delimited territory—and hence the subdivision of a population into plural hierarchised populations—is not explored enough by Foucault. This has consequences for his analysis of racism, as I discuss in the next section. What also remains underexplored is the degree to which governmental action that purports to enhance the health of a national population is actually aimed at doing so, with a strategy founded on medical evidence, or whether this aim is confounded by other motivations of neoliberal governmental rationality.

Concerning the biological policing of HIV described here—which I call 'biopolicing'—it is notable that, among the medical community, the dominant cause for the spread of HIV in Europe, as well as in Greece, is considered to be intravenous drug use, and harm reduction strategies are considered more effective in controlling the virus than arrests and quarantining (*Lancet* 2013). Indeed, the Hellenic Centre for Disease Control and Prevention (HCDCP) (2011) identified a major HIV 'epidemic' in 2011 specifically among intravenous drug users (a 63 percent increase) and not within other populations. The Greek government's legislative regulation that permitted the forcible HIV testing and their imprisonment and stigmatisation of immigrants and sex workers were condemned in major medical journals (Sarafis, Igoumenidis and Tsounis 2013; *Lancet* 2013). There is also evidence from several countries that economic crises are associated with HIV epidemics among intravenous drug users (Friedman, Rossi and Braine 2009). In Greece, health budget cuts of 30 percent in the first quarter of 2011 (Simou and Koutsogeorgou 2014) caused decreased coverage of harm reduction services and falls in the number of syringes (–10.3 percent) and condoms (–23.9 percent) exchanged (Fotiou et al. 2012).

Yet, undeniably, the government carried out arrests and forced testing with some scientific support, via molecular epidemiology analysis, for the notion that immigration was associated with the epidemic (Paraskevis et al. 2011). An 'expert report' of the Greek National REITOX Focal Point (2011) had also speculatively associated 'illegal immigrants' (p. 4) and opioid-dependent sex workers with the epidemic, without providing any concrete evidence. Although none of these reports recommended arrests, forced testing or quarantining, the government received the support of HCDCP medical professionals who carried out HIV testing without the patients' consultation or consent, against international medical guidelines (*Lancet* 2013, 102).

It is not sufficient to suggest that individuals in the medical profession and the HCDCP became involved in a politically motivated campaign, as they later admitted, offering belated apologies to the jailed women (Ī Augī 2016)—but not to the immigrants whose residences were evacuated. These actions ought to be placed in a broader context of how neoliberal governmentality has, in recent years, shifted its focus from a 'rights-based' management of public health to a risk-management approach that fits within a *dispositif* of security (Foucault 2009 [1978], 16–20). As Julia Smith (2016) notes in a survey of recent governmental responses to HIV in the EU, including the case of the Loverdos regulation in Greece, the passage from 'rights' to 'risk' entails a passage from the universal provision of health care that attempts to prevent social marginalisation to the separation of populations between those who are the 'threat' and do not have access to free health care, and those who are to be protected on the basis of their citizenship and economic status. These divisions were deepened in the crisis, when the lack of access to free health care became a problem not only for immigrants, who had access only as asylum seekers or refugees (a status extremely difficult to attain) but also for the unemployed.[5] Until 2016, access to free health care for the unemployed depended on the level of their national insurance contributions.[6] Mass unemployment and the expansion of precarious and uninsured employment meant that large numbers of citizens lost access to free health care.

The *dispositif* of security here takes a specific shape, justified partially by cost-cutting considerations, under which the identification of 'at risk' populations is no longer on the basis of the risks they face but on the basis on the risks they pose to the populations whose health the government prioritises—in this case, Greek citizens and specifically married men. The 'threatening' populations are treated as not possessing the same rights (to privacy, health care, habeas corpus) as those who belong to the prioritised citizen categories. The very conception of 'risk' is thus founded on the othering of populations whose illness is only of interest to the government insofar as it is infectious. These patients are quarantined in prisons and detention centres where they presumably pose no risk to 'Greek society', but where they face higher health risks and endanger other detainees. The 'threat' is thus no longer the virus itself or the failures of sanitation, housing provision, public health care and harm reduction programmes—which are justified by the logic of economic rationalisation—but 'at risk' populations themselves, whose illness is criminalised. This risk-management approach defines the research questions that come to be posed by a section of medical science: the geographic profiling of the 'origins' of infections by Paraskevis et al. (2011), and their attribution of the problem to immigration, is politically motivated insofar as citizenship itself is political. We are thus not simply seeing a cover-up of public health cuts. Instead, a specific logic of managing public health becomes merged with the policing of borders and sex work, founded as it is on the hierarchisation of populations.

In the field of policing itself, the prime aim becomes not crime in general but the crime of profiled populations. Racial profiling, now involving extensive DNA databases (Roberts 2011), has been much discussed for its role in the police murders of African Americans in the United States. In the medical field, racial profiling follows a similar logic, based on rationalisation and economic efficiency: resources are targeted at controlling and incarcerating a population presumed to pose a risk to the white 'general population'. In the Greek case, similarly, immigrants are the primary population targeted in high-profile public-order operations like Xenios Zeus. Yet the concern is not only crime itself but the 'citizens' sense of safety' and the 'image' of city centres (Dendias 2013, 1), which affect market operations. This type of profiling is founded less on statistics than on heuristic impressions of what might improve a city's image from the perspective of 'the citizen', leading to the targeting of the homeless, drug users, and people of nonheteronormative/cis genders (see also Human Rights Watch 2015). This maintains an exclusionary aesthetic norm in public space that mirrors the way in which hierarchical relations of class, race, heteronormativity and gender are mapped onto who 'the citizen' is presumed to be.

Thus, against Foucault's assertion in his 17 March 1976 lecture (2003), the securitisation of health does not take as its object a single population that is the human species as a whole. Instead, it is founded upon a prior separation of human life (multiple populations) between national borders, as well as on the basis of pre-existing social hierarchies within and between nation-states. Moreover, its motivation is not so much the enhancement of the (national) population—a long-running aim of existing rights-based healthcare structures—but it forms part of strategies aimed at decreasing state expenditure and increasing its economic efficiency.

EXCEPTION AND IMMUNISATION IN NEOLIBERAL CRISIS MANAGEMENT

Still, even though this specific manifestation of the security *dispositif* takes place in the context of a neoliberal restructuring, its disregard of rights and its deepening of racialised and gendered segregations raises the question of whether it represents a tendency towards an antiliberal, or even fascistic, departure away from the neoliberal paradigm. The question of continuity or discontinuity between modern liberal biopolitics and totalitarianism is one that has concerned Foucault, as well as Giorgio Agamben and Roberto Esposito, all of whom have used the example of the Nazi regime to explore this question. Below, I examine the relevance of this work for understanding the operation of biopower in the examples discussed here. I identify some potential blind spots with regards to understanding neoliberal governmentality, which require further exploration and synthesis with elements taken from other theoretical traditions, namely, Marxism and psychoanalysis.

Foucault's analysis of biopolitics poses the question of totalitarian elements within liberal democracies when he identifies the conditions of possibility of the Nazi regime in modern biopolitics: '[f]or the first time in history, the possibilities of the social sciences are made known, and at once it becomes possible both to protect life and to authorize a holocaust' (cited in Agamben 1998, 10). What makes both outcomes possible is the common mission to 'control the random element in biological processes' (Foucault 2003, 246). As is well established, eugenics was not an innovation of the Nazi regime, but it was a scientific and political orientation that was taken to be a progressive intervention to enhance the life and health of the population, and even women's freedoms, across the political spectrum (Deutscher 2010). This introduction of a caesura into the biological domain, between what must live and what must die, for the purpose of enhancing the life and health of the nation, becomes, in Nazi biopolitics, the logic that separates groups within a population, as well as enables war by identifying the enemy population and justifying its death (Foucault 2003, 254–62). This is, for Foucault, the definition of racism.

Yet, the weakness of this definition soon becomes evident. It both lacks specificity and fails to identify the racial politics of neoliberal *dispositifs* of security. In tracing the origins and logics of neoliberal biopolitics, Foucault (2009, 106–17) notes how market freedom was promoted by German ordoliberals in combative juxtaposition with the Nazi regime, to whose economic planning they attributed its deadly outcomes, and which they categorised alongside welfarist and social-democratic states. Foucault is not convinced by this identification of the Nazi regime with welfarist social democracy. However, in his own analysis of racism as the necessary precondition for waging war on enemies, he comes to equate not only Nazi but also socialist, communist and anarchist combative politics with racism:

> Whenever, on the other hand, socialism has been forced to stress the problem of struggle, the struggle against the enemy, of the elimination of the enemy within capitalist society itself, and when, therefore, it has had to think about the physical confrontation with the class enemy in capitalist society, racism does raise its head, because it is the only way in which socialist thought, which is after all very much bound up with the themes of biopower, can rationalize the murder of its enemies. (Foucault 2003, 62)

Here racism loses all specificity and becomes nothing but a means to identify an enemy, regardless of the logic of this identification. Indeed, the designation of the 'enemy' defines who must live or die under conditions of war, and hence it falls within the domain of biopower. Yet, there is a substantial difference between (a) the conception of a biological threat to the body of the nation, its presumed biological and cultural *ethnos*, and its familial genealogies, and (b) the struggle that takes place against an agent who is only identified by their position of power within a political-economic structure, and not by their biological or genealogical characteristics. This distinction has consequences for analysing contemporary governmental practices, because the centrality of ethnic genealogy and the patriarchal family in racist-nationalist thought is one that also opens up the questions of gender and sexuality (see Weinbaum 2004). It also recognises that racialised groups are often forced to defend themselves physically against political organisations and state agents who perpetuate their violent racialisation and often their death. Finally, it underlines the difference between sovereign biopower and the violence that can potentially be perpetrated in the course of resistance to it—notwithstanding the complexity of the relationship between power and resistance that Foucault (1982) has identified.

Giorgio Agamben (1998) has pursued further the problematic of sovereignty, emphasising the necessary overlap between biopolitical and juridical sovereign power against Foucault's analytical separation. In his analysis, sovereign power produces the citizen both as a political and a biological subject. As a political subject, the citizen has a political life ('bios') endowed with political rights. Yet both sovereignty and the citizen are constituted by the state of exception. The exception does not only demarcate the outsider—the noncitizen who lacks rights by definition—but also the citizen whose rights can be rendered inoperative. The 'sovereign ban' under a state of exception has the capacity to render any citizen no longer a political being, but a 'bare' life, a mere biological body, that can be killed at the threshold of the law. With this definitional distinction Agamben draws parallels between the 'thanatopolitics' of Nazi concentration camps and the function of the exception in contemporary liberal democratic nation-states. In both, exceptional state violence and extralegal forms of social control and population management, which reduce citizens to their bodily characteristics, are implied in the very possibility of sovereignty and citizenship.

Agamben (2005, 18) identifies strong continuities between totalitarian and liberal democratic state practices, and draws special attention to the increase in the powers of the executive in liberal democracies over at least the past half-century. In our case, the operation of the 'exceptional' state, through executive ministerial orders, produces populations that not only fall into a legal exception and lack rights but also come to be victimised by an uncontrolled repressive state apparatus that collaborates with neo-Nazi vigilantes. Under the conditions of the exception, sovereign biopower becomes terror through allowing citizens to partake in the cleansing out of life that, devoid of rights, no longer belongs to the body of the citizens. However, Agamben's analysis is unable to explain which bodies fall into the sovereign exception and for what purpose. The rationality of modern biopower, on whose growth Foucault places so much importance, is de-emphasised to highlight the arbitrary element in sovereign power. But there are ongoing historical patterns which suggest that specific social categories, defined by gender, race and class, tend to fall into the realm of exception. There are also historical shifts in the degree to which the state of exception is exercised and the forms it takes. There is thus more that determines the exercise of sovereign biopower, as well as more contingency, than Agamben allows for.

A more historicised account of the move from modern biopolitics to Nazi thanatopolitics is provided by Roberto Esposito, who analyses the modern liberal nation-state as constituted by a 'paradigm of immunisation' (2008, 45–77). Esposito enquires into the dialectic between community and immunity established by modern sovereignty, whereby the individual is immunised from owing any gift or debt (*munus*) to the community precisely through his rights of property and liberty.[7] In this way, he identifies a shared biopolitical lexicon between the immunisation of the liberal individual and the immunisation of the community in its most self-destructive, 'auto-immune' form in Nazi biopolitics. Esposito highlights the medicalised character of Nazi government (the omnipresence of doctors in decision-making) and the intended therapeutic and protective function of the camps for the German *genos* from the ostensible threat of Jewish degeneracy. Here, the absolute immunisation of the national body (making live) from the 'threat' of foreign becomes auto-immune: it destroys the national community itself, by coming to justify the mass murder of 'life unworthy of life' (thanatopolitics), both within and beyond what are taken to be the borders of the community. Esposito criticises the centrality of internal and external immunisation in the modern constitution of the community.

Esposito's work is relevant to the questions posed here in several respects. First, he points out the enduring political and etymological link between birth (*nascita*) and 'nation' revealing its inherently racial meaning (2008, 169). Thus, racism is not an ideological aberration but remains at the centre of modern democratic nation-states. Second, the immunisation paradigm draws attention to the intimate links between the racialised, sexualised and gendered selectivity of modern biopower through the problem of generational reproduction. Women's reproductive role and thus their pivotal position in the control of health and racialised-genealogical dimensions of populations has historically rendered their bodies into targets of biopolitical control—most paradigmatically in eugenics. This theme has re-emerged in contemporary 'neoreactionary' racist discourses that merge narratives of technologically enhanced white genetic and cultural superiority (Haider 2017). Yet, beyond the reproductive activity or potential of the bodies concerned, Esposito (2008, 172–73) notes the gendered biopolitical meaning of *fraternity* on which the modern concept of democracy rests. 'If it is true that democracy is often referred to the idea of brotherhood, that is because democracy, like all modern political concepts, rests on a naturalistic, ethnocentric, and androcentric framework that has never been fully interrogated' (173).

Esposito's observation strengthens the argument put forward here that, for liberal states, 'population' has never historically referred merely to a biological body of humanity defined outside its belonging to a nation-state. Rather, it is tied up with the historical meanings and separations of citizenship and thus the prioritisation of male citizens. Similarly, the gendered dimension of biopolicing, in our case, rests not so much on the reproductive capacity of the sex worker but on her sexual activity alone. The selectivity of immunisation, which entails the protection of the male citizen's body from the female, constitutes the sex worker as a threat to the *fraternal* democratic national body. This observation certainly does not fully explain the dynamic links between sex work, immunisation and gender. The illegality of sex work and the cultural and psychic aspects of its history also come into play, and they are briefly explored in the following. Yet it does reveal the congruence between this type of hierarchisation of bodies and liberal democracy.

This is Esposito's most useful contribution. He detects the links between nation, biology, race and sexuality as not only characteristic of the Nazi regime but also of liberal-democratic regimes. To the degree that the figure of the 'political body' remains central to the constitution of the liberal nation-state, the immunitary paradigm of the 'self-preservation of the political organism as a whole' (Esposito 2008, 158) prevails. There would appear to be a common aim, therefore, behind the government's failure to distinguish among the levels of risk carried by a range of illnesses and its construction of racialised and gendered biological threats against which it develops violent practices of immunisation. These practices can be understood as a paroxysmal—and futile—attempt to reunite a disintegrating national community and so strengthen the entire 'political organism'—the civil society and the state—along the lines of the immunitary paradigm. Yet it is important at this stage to return to the question of neoliberalism. If these forms of biopolitics are characteristic of modernity as claimed by Esposito, what is distinctive about the neoliberal *dispositif* of security?

The abstract political forms of modernity cannot be understood unless we also pay attention to their concrete historical forms, which are implicated with capitalist social relations, colonialism and the formation of nation-states. If there is one point on which Marxist and Foucauldian analyses agree, it is that the requirements of capitalist reproduction play a central role in governmental decisions to enhance or control the number, health and productive capacities of populations. Yet the imperatives of capitalist reproduction and the phases of capitalist development over the course of the 'modern' era are not taken into account sufficiently by Foucault, and even less so by Agamben and Esposito. The role of social struggle in shaping the exclusionary histories of nation-state formation, citizenship and suffrage is also left aside, yet it is part of what helps us define the specificity of biopolitics in the neoliberal era.

Social struggles within liberal democracies over the twentieth century effected shifts in the character of liberal law and politics through accommodations in liberal legal systems that went beyond formal equality. With the neoliberal shift, these accommodations, which often took the form of redistributive measures such as welfare benefits, universal health care, free childcare, alimony and free legal support, have been seen as involving excessive state power that limits market freedom and were even associated with totalitarianism. Yet the market freedom that has been established through the neoliberal restructurings of the past forty years amounts to another mode of governmentality, a financial governmentality that imposes a form of economic rationality that can come into contradiction with prior governmental aims such as the enhancement of the population. The latter is no longer an aim in itself but ought to be the end result of a freely functioning market in which the state only intervenes to facilitate. Here, we should distinguish between an exploration of debt as a form of neoliberal subjectivity (Lazzarato 2012) and the analysis of market-led governmentality, which incorporates the role of finance and neoliberal ideology. I focus on the latter in the following text, because it is directly relevant to the production and function of biopolitical caesuras that distinguish populations, at a time when the neoliberal economic paradigm faces a crisis.

According to the analysis of Sotiropoulos, Milios and Lapatsioras (2013), in financialised capitalism, states face restrictions to how they carry out budgetary management because, apart from being political entities, they are also economic entities. The market value of their sovereign bonds fluctuates depending on the health of their finances. This is another facet of risk management from the point of view of a multitude of investors, which, in turn, affects states' ability to borrow and fund their expenses. After the 2008 financial crisis, it became evident that national states are subject to a financial type of governmentality, which radically limits their autonomy in governing a territory and ensuring capitalist accumulation within it. In the EU, the interdependence among states produced by the common currency has also meant that states in the Eurozone cannot make independent decisions on managing and stabilising the local economy. The role of the IMF and the EU, as supranational institutions who oversee and regulate the European and global economies, is pivotal here, in prescribing policies that aim to guarantee continued accumulation and minimise the devaluation of financial capital. This can draw the national state, particularly a deeply indebted state, into a severe political crisis and produce problems of legitimation, because the state no longer has the flexibility to manipulate policy in order to achieve a balance between ensuring the servicing of debt, continuing accumulation, and securing the subsistence of working and nonworking populations. While the national state and its sovereignty *appear* as the fundamental bulwarks against financial governmentality, in practice state power is used to impose the imperatives of balanced state finances and flexible labour markets regardless of the human cost. The impersonal 'punishment' of states for sovereign debt and insufficient 'reforms' is transferred to individuals, who are impacted differentially depending on their economic position. The punishment is particularly severe when demand for labour power is low and increasing proportions of the population are cast off as surplus to the reproduction of capital.

The subjection of the state to the abstract governmentality of the market thus deepens the already existing social separations, both because the market produces caesuras within the population and because the life of those who suffer market 'failure' is no longer a concern. To add to Agamben's terminology, under the governmentality of free market competition, it is not only the form of right (sovereignty/exception) but also the content of rights (the primacy of negative economic freedom) that renders the distinction between bare life and political life problematic. At the limit, under the terms of contract, freedom, property and equality before the law, the biological life of proletarian citizens is indifferent. While Esposito does comment on the immunising function of the right to property from the obligation of *munus* (gift or debt) to the community, he similarly neglects to discuss the thanatopolitics of the *other* body constituted by property and liberty, namely that of the propertyless proletarian, who stands, formally, under the equal 'protection' of the law. Nobody has expressed this better than Evegniĭ Pashukanis: 'The personality of the proletarian is "in principle equal" to the personality of a capitalist; this finds its expression in the fact of the "free" contract of employment. But for the proletarian this very "material freedom" means the possibility of quietly dying of starvation' (1980, 105).

It is relevant here to highlight the continuities and contradictions between neoliberal theory and the meritocratic ideology of protestant theodicy mobilised to legitimise this order of things. For the latter, the laws of the market have expelled those less competent from the labour contract, and have separated the healthy social body from the diseased or incompetent. The life of the diseased body is either indifferent or it can even become undesirable, and falls into the realm of the exception, so that it does not threaten the economic 'health' of the body of the nation by its existence in metropolitan city centres. This power dynamic is consistent with the immunitary paradigm described by Esposito: the (economic) community is taken to be protecting itself through the immunising function of the market, which expels weak elements and strengthens the rest of the economy. On the other hand, in Hayek's work (1976) we find the admission that the market does not reward according to any principle at all, and cannot be made into an instrument of 'collective value or social justice' (Plant 2004, 36). This means that the elements (and human beings) the market mechanism expels do not correspond with what or who a community may regard as worthy: it is an immunitary mechanism that follows its own logic. Hayek, however, considers 'whether without such partly erroneous beliefs the large numbers will tolerate actual differences in rewards which will be based only partly on achievement and partly on mere chance' (1976, 74). He recognises the need of such beliefs for social legitimation. Of course, 'chance' is very far from describing the systematic processes through which market competition disadvantages those who are already disadvantaged. If, however, the social legitimation of the market depends on its ability to separate between the capable and healthy social body and the incapable and undeserving, we can see how the reinforcement and moral justification of existing social separations can be a neoliberal governmental technique of self-legitimation. This is particularly pressing for the neoliberal state, whose sovereignty is subservient to the market. In a moment of economic crisis, which, as discussed, is also a political crisis and a crisis of the nation-state's sovereignty, power can be re-established through the exercise of an exception, whereby *specific* citizens and noncitizens become treated as contagious threats to the political body as a whole. The caesurae that define those bodies are the sediments of historical economic, racialised and gendered social divisions, which are deepened by the abolition of the moderating accommodations of social democracy.

Thus, through this form of crisis management, economic nonintegration increasingly means a political nonintegration, and the historical racialised, gendered and class exclusions of citizenship become again effective. It becomes explicit that the citizen is not the undefined body of liberal legal equality, but a hierarchy is reinforced, through which the body politic comes to be represented primarily by the male body, which stands in for the mass of households that are taken to comprise 'Greek society'. That body bears no responsibilities as a sex work client or virus carrier but is only to be protected from the potentially diseased bodies of sex workers.

But while the realm of legal exclusion broadens for the domestic surplus population, citizenship has been also reaffirmed through an even deeper and more violent political exclusion of migrant proletarians, especially given that the cheapness of their labour is premised precisely on migrants' irregularity, which deprives them from any rights or protection. This has played out in the preference for 'fencing' as opposed to 'gatekeeping' external and internal immigration controls (Triandafyllidou and Ambrosini 2011). In the policing of its eastern border with Turkey, Greece strengthened its fencing practices with EU funding in 2011 and with the contribution of Frontex forces. Yet a great many immigrants still made it into the country,[8] who were again managed exclusively through spectacular 'fencing' measures such the Hospitable Zeus operation. On the contrary, no 'gatekeeping' measures have been in place or actively enforced that would control immigrants' informal employment or to prosecute their employers (ibid.). The shootings of immigrant agricultural workers by their supervisors in Nea Manolada, for which their employer was acquitted (Smith 2014), and similar cases elsewhere in the Peloponnese are emblematic of this policy. Immigrant workers are exposed to employers who have alternately mobilised GD or the police to terrorise their workforce into tolerating unpaid work and degrading living conditions. Thus, the citizen is absolved of any crimes against a particular category of noncitizens, proletarian immigrant workers, in the interests of creating a hyperexploitable, racialised, and continually persecuted proletarian segment in the Greek labour market. This retains the gap between the prices of the local population's labour power and that of the immigrant, while devaluing them both, producing optimal labour market conditions for investors.

We see then how national borders are reinforced at the level of proletarian populations at the same time as they are eroded at the level of the global market and economic sovereignty. The nationalist discourse of the Greek governments of 2012–2014 was not anachronistic but was related to the necessities of capitalist reproduction, at a time when neoliberal crisis management strategies produce populations that are surplus to capitalist accumulation. Its racist and sexist discourse was not only motivated by the imperatives of enhancing the life of the population and defining enemies (Foucault), of protecting the national community through immunisation (Esposito), or of establishing sovereign biopower (Agamben). In contrast to fascist regimes of the past, this was a strategy coloured by the attempt to gain social legitimation *for* neoliberal governance, whose imperative is increasing exposure to the immunising mechanisms of the market.

BIOPOLITICS, ABJECTION AND FANTASY

It is then clear that the governmental operations discussed here—the arrests, forced testing and public exposure of HIV-positive women; the evacuations, arrests and quarantining of immigrants; and the Hospitable Zeus operation—had simultaneously a functional and a performative-communicative role. It was not enough for the policing operations to simply happen: they happened in front of a nationwide audience. It is in the eyes and ears of this audience that neoliberal governmental logic needed to be legitimised, through the stigmatisation of specific populations.

To discuss how, through the demands of neoliberal social legitimation, already marginalised populations become additionally stigmatised, Imogen Tyler has recently developed the biopolitical and psychosocial concept of 'social abjection' (2013, 46). Drawing on Bataille's definition of abjection as 'the imperative force of sovereignty, a founding exclusion which constitutes a part of the population as moral outcasts' (19), she has analysed political rhetoric that cultivates social disgust towards unemployed, impoverished and migrant populations, and especially women, through narratives of hyperfertility and slothfulness. The production of the abject moral outcast then plays the role of enlisting social consent for neoliberal reforms, which, as discussed above, are legitimised through the myth of the market as a reward mechanism for the 'deserving'. Similarly, in Loverdos's public health operation, disgust was stirred by presenting the biopoliced population as objects posing a visceral threat to the 'mainstream' body of citizens. They were not subjects with whom there could be a dialogue but infectious bodies to be prevented from coming into contact with others: they became abject.

Beyond simply documenting victimisation, Tyler's work in *Revolting Subjects* is also devoted to revealing the ability of abjected subjects to resist abjection. She rejects the psychoanalytic theory of Julia Kristeva (1982), which has accompanied much of the term's use in theorising racism and misogyny, for its inability to recognise the agency of resistance. Yet this focus on the relationship between governmental biopower and resistance from below leaves out of the picture the role of the spectator who consents to this governmental action. Her work does not ask why the spectacle of the other as a viral threat to family and property is politically effective upon the intended audience, or, at least, why there is the expectation that it will be. This is an important question if we seek to understand why and how the spectacular aspect was intrinsic to the policing operations described in this article.

We have seen that, through these governmental practices, the citizen is not treated as an empty vessel of liberal legal equality but a gendered hierarchy among citizens based on the *household* is reinforced. 'Household' has the multiple meanings of ethnic genealogy, fatherland, property and patriarchal family, which are far from emotionally indifferent. The purported rationality of risk management and law enforcement contains a substantial element of irrationality and fantasy. Those interpellated by the designation of 'family men'—whose voices were privileged in the media at the time—were said to be threatened by a bigger danger than poverty and unemployment: their fatal viral infection by a woman. Despite evidence that it is typically clients who pressurise sex workers not to use condoms (Sarafis et al. 2013), sex workers' sexual encounters were treated as their sole agency and responsibility, and their infected bodies were imagined as weapons. In a stereotypical inversion of the gendered passive/active binary, here the man was fantasised as the passive victim of a seductive and deadly female figure who manipulates the male sex drive and threatens the lives of families.

We are here in the realm of male fantasy. This type of narrative was the norm for hundreds of years in the scientific and policy discourse on syphilis, which presumed prostitutes and foreigners to be the 'source' of the disease

(Grünpeck 1497). The fact that the disease was also transmitted from men to women was not of particular interest, neither was the possibility of transmitting it to citizens of other countries (Association for Promoting the Extension of the Contagious Diseases Act 1868). Thus, the concern is not the disease as such but its invasion into the male body, and, by extension, into his household. The powerful household analogy of 'host' and 'visitor' by which national identification also operates opens the way for analysing the question of borders in the terms of individual and collective bodily integrity, which also explains the fascination with myths and incidents involving rape or sexual assault by immigrant men on 'native' women.

The fear of invasion, and of the threats to the borders of identity, is a central theme of the psychoanalytic concept of the abject, as examined by Julia Kristeva (1982): it is 'not lack of cleanliness or health that causes abjection but what disturbs identity, system, order' (4). Thus, what threatens is not illness as such but the fact that 'borders, positions, rules' (4) are violated. Abjection, from this psychoanalytic perspective, is not simply synonymous with disgust, but it is founded on a theory of ego formation and of the fear of its dissolution. The ability to separate the self from the maternal and recognise her as other, and simultaneously to reject her as part of the self, creates self-coherence. At the same time, this self-coherence depends on entering the symbolic order, which is social and historical, and is permeated by the sediments of the dynamics of patriarchy, capitalism and colonialism. Keeping that in view, it is possible to make careful use of psychoanalytic theory both through historicising it as a form of knowledge-power shaped by colonialism (see Khanna 2003) and through de-universalising the forms of subjectivity and individuality it describes.

Fanon (1967) was the first to dissect, using psychoanalytic tools, how white European individuality was psychically shaped by colonialism. Although 'abjection' was not a term he used, his work shows that the association of white subjectivity with rationality and civilisation is founded on caricaturing the other and projecting onto her everything that this symbolic order rejects: weakness, femininity, primitiveness, illness. Of course, this is accompanied by immense physical, economic and psychic violence. Yet, the other, while victimised, becomes a threat at the 'border' moment of indistinction and anxiety, in which identity is not clearly circumscribed and might be threatened by invasion: the moment of abjection.

The historicity of this symbolic order would also guide us to recognise what Kristeva calls the 'necessity' of 'matricide' (1989, 27)—the misogyny at the core of abjection—as only necessary within a definite historical-social configuration. The patriarchal culture that associates masculinity with individual independence and autonomy comes to interact, in a volatile way, with the independence, competitive 'capability' and self-reliance valued by (neo-)liberal ideology. The contradiction between the myth of the market as a reward mechanism and the actuality of its arbitrary immunising logic is a source of social and psychic tension. The feminised element within the self that is at risk of 'failure'—and here the significance of the passage from 'rights' to 'risk' in the security *dispositif* becomes evident—is feared and ought to be expelled, in the same way as the market's immunising mechanism expels those no longer essential to the reproduction of capital.

The operation of abjection in this sense demonstrates how the dynamics of patriarchal misogyny, Eurocentric racism and the stigmatisation of the poor intersect not only in the practices and ideologies promoted by neoliberal and increasingly authoritarian states but also within the subjects produced in such a society. The fear of the marginalised other is also a denial of the self as potentially in need of others, and thus dependent and emasculated. In Kristeva's words, 'There is nothing like the abjection of self to show that all abjection is in fact recognition of the *want* on which any being, meaning, language, or desire is founded' (1982, 5).

The psychosocial conjuncture of the concern with borders and sexuality can then be linked to the historical form of immunised individuality produced in modern liberal societies, and the tensions this form of individuality is subject to under a neoliberal model in crisis. It is worth asking what might be the psychic parallel here with Esposito's imagining of an 'affirmative biopolitics'—inspired, similarly to Kristeva and Tyler, by Bataille—that recognises flesh as 'constitutively plural, multiple, and deformed' (Esposito and Campbell 2006, 2) and destabilises the boundness of individual life. It could entail a form of psychic subjectivity and culture that does not dread a degree of fusion, contagion or interdependence with the other. Yet that could not be achieved under a global social system whose measure of 'health' is founded on the immunising function of the capitalist market and its legitimising ideology.

CONCLUSION

I have discussed how the security *dispositif* of contemporary neoliberal crisis management meets with the active separation, targeting and quarantining of populations that are constructed as posing 'threats' to the prioritised body of citizens. At least in the cases examined, set in Greece in 2012–2013, this was clearly not only a risk management technique but also one that could be used to legitimise the economic immunising mechanism of the market, through its myth as a mechanism that meritocratically separates the healthy body of citizens from the deficient or undeserving. This tool became relevant at a moment of deep capitalist crisis and of the neoliberal model of government, which yields state sovereignty to the facilitation of market freedom and financial governmentality. The populations demarcated by the caesurae of this market process are not the result of a random operation, but they are defined by the histories of exclusion that have shaped the racialised and gendered social hierarchies internal to the meanings of Western liberal citizenship. However, as a governmental technique that sought to have affective power as a spectacle, its operations of social abjection depended on the invocation of fantasies of invasion—into the land, property, household or body of the citizen.

The biopolicing practices of the Greek governments of 2012–2013 are characteristic of the contemporary political context, in which neoliberal politics appears to be giving way to nationalist authoritarianism. Yet, even though the latter appears as a resistance to the former, the present analysis shows that the modes of legitimation of neoliberal governance in the current crisis have involved similar modes of thanatopolitical immunisation of the body politic through placing populations that constitute 'threats' under a state of exception. The continuities between liberal biopolitics and totalitarian modes of government pointed out by Foucault, Agamben and Esposito are thus evident in the present conjuncture. In contrast to fascist regimes of the past, however, what Esposito might call governmental practices of immunisation are not being carried out to enhance the life of the *genus*. Instead, their role, and particularly their spectacular, stigmatising dimension, is to legitimise exposure to the imperatives of the market: the abstract immunising function of the market is presented as containing a moral dimension that is consistent with the protection of the health and security of the national community. Despite this effort, however, the contradictions between the two seep through. In Greece, a left-wing type of nationalism (via SYRIZA and its anti-EU offshoots in 2015), and nationalist authoritarianism elsewhere, have legitimised discourses that contrast the national economy to the global market. But this 'conflict' between (neo-)liberalism and nationalism is really a conflict between two forms of immunisation, both of which understand 'population' exclusively through the (gendered and racialised) genealogical demarcations and international hierarchies of nationality and citizenship. Deeply questioning the naturalised assumptions of these separations is thus a key precondition for a contemporary emancipatory project that can challenge both the brutality of the financialised capitalist economy and the abjection of self and other that the nationally circumscribed community and economy entails.

NOTES

1. 4046/2012 AK. Specifically, among other labour 'reforms', the second memorandum's package of reforms reduced the minimum wage by 22 percent and by 35

percent for those under age twenty-five; limited the validity of collective agreements, so that wages would fall to the minimum after their expiry; and froze wages until unemployment fell below 10 percent.

2. G. Y. 39a, FEK B′ 1002/02.04.2012.

3. See the 2013 documentary *Ruins: Chronicle of an HIV Witch-Hunt* by Zoe Mavroudi about the women's arrests, their public shaming, and the sexist discourse that surrounded the entire case: http://ruins-documentary.com.

4. In Ancient Greek religion, the epithet 'xenios' designated Zeus as the patron of hospitality (*xenia*) and guests, and the avenger of wrongs done to strangers.

5. Greece's first instance asylum recognition rate in 2012 was as low as 0.84 percent (Bitoulas 2013).

6. The problem of health-care access was partially addressed by the SYRIZA government in 2016 with a new law (N. 4368/2016 AK, Ar. 33), which extended free access to the uninsured unemployed and to all documented immigrants, as well as those in camps and detention centres. Only recognised refugees and asylum seekers had free access previously. Immigrants without documents continue not to have access to free health care, but they can access primary and emergency services.

7. The generic 'he' is consciously used here, since, at the time of its establishment, modern democracy afforded rights only to propertied men.

8. According to Eurostat data, 72,420 were found to be illegally resident in Greece in 2012, all of whom were ordered to leave, while only around 15,746 of those were deported. A very small proportion, 9,575, were able to make asylum applications. An enormous proportion of migrants remain in the country clandestinely, constantly at risk of abusive encounters with police.

BIBLIOGRAPHY

Agamben, Giorgio. 1998. *Homo Sacer: Sovereign Power and Bare Life*. Palo Alto, CA: Stanford University Press.

Agamben, Giorgio. 2005. *State of Exception*. Chicago: University of Chicago Press.

Association for Promoting the Extension of the Contagious Diseases Act. 1868. *Report on the Extent of Venereal Disease, on the Operation of the Contagious Diseases Act, and the Means of Checking Contagion*. London: H. Baillière.

Bitoulas, Alexandros. 2013, May. "Asylum applicants and first instance decisions on asylum applications: 2012." *Eurostat: Data in Focus*.

Chatzīgeōrgiou, Arīs, and Ntani Vergou. 2011. "Loverdos: Kindynos Gia Tīn Ellīnikī Oikogeneia—'Na Apelathoún Oi Ieródoules Foreís Tou AIDS'" [Loverdos: A Danger to the Greek Family—"Deport HIV-Positive Sex Workers"]. *Eleutherotypía*, 16 December.

Dendias, Nikolaos. 2013, June 28. *Parliamentary Answer by the Ministry of Public Order and Protection of the Citizen*. No. 7017/4/16499.

Deutscher, Penelope. 2010. "Reproductive Politics, Biopolitics and Auto-Immunity: From Foucault to Esposito." *Journal of Bioethical Inquiry* 7 (2): 217–26.

DOL Newsroom. 2014. "Antinazi Zone: Meiōthīkan Oi Ratsistikés Epithéseis, Auxīthīkan Oi Omofovikés" [Racist Attacks Decreased, Homophobic Increased]. *In.gr*, 31 October.
eKathimerini. 2012, January 4. "Compulsory Health Checks for Migrants," http://www.ekathimerini.com/140488/article/ekathimerini/news/compulsory-health-checks-for-migrants.
Eleutherotypía. 2013a. "Athōes Oi Orothetikés Katópin Diapómpeusīs" [HIV-Positive Women Acquitted after Public Shaming], 17 January.
Eleutherotypía. 2013b. "Aúxīsī Stis Omofovikés Epithéseis Me Tīn Ánodo Tīs Akrodexiás" [Increase in Homophobic Assaults with the Rise of the Far Right], 21 January. http://www.enet.gr/?i=issue.el.home&date=21/01/2013&id=337482. Accessed 31 August 2017.
Esposito, Roberto. 2008. *Bíos: Biopolitics and Philosophy*. Minneapolis: University of Minnesota Press.
Esposito, Roberto, and Timothy C. Campbell. 2006. 'Interview'. Translated by Anna Paparcone. *Diacritics* 36 (2): 49–56.
Fanon, Frantz. 1967. *Black Skin, White Masks*. New York: Grove Press.
Fōtiadī, Iōanna. 2014. "Plīthaínoun Oi Omofovikés Epithéseis Sto Kéntro Tīs Athīnas" [Soaring Homophobic Assaults in Athens City Centre]. *Kathīmerinī*, 30 August.
Fotiou, Anastasios, Katernina Micha, Dimitris Paraskevis, Manina Terzidou, Melpomeni M. Malliori, and Angelos Hatzakis. 2012. "HIV Outbreak among Injecting Drug Users in Greece." Athens: European Monitoring Centre for Drugs and Drug Addiction. http://www.emcdda.europa.eu/attachements.cfm/att_191984_EN_HIV_update_Greece_2012.pdf. Accessed 31 August 2017.
Foucault, Michel. 1982. "The Subject and Power." *Critical Inquiry* 8 (4): 777–95.
Foucault, Michel. 2003. *Society Must Be Defended: Lectures at the Collège de France, 1975–76*. New York: Picador.
Foucault, Michel. 2009. *Security, Territory, Population: Lectures at the Collège de France 1977–78*. Lectures at the Collège de France. Basingstoke, UK: Palgrave Macmillan.
Friedman, Samuel R., Diana Rossi, and Naomi Braine. 2009. "Theorizing 'Big Events' as a Potential Risk Environment for Drug Use, Drug-Related Harm and HIV Epidemic Outbreaks." *International Journal of Drug Policy* 20 (3): 283–91.
Fyntanídou, Elena. 2012. "Epidīmía Tou Ioú HIV Sto Kéntro Tīs Athīnas" [HIV Epidemic in the Centre of Athens]. *To Vīma*, 5 June.
Galanoú, Marína. 2013. "Ótan Ta Trans Átoma Proságontai, Gia Na 'Veltiōtheí Ī Eikóna Tīs Pólīs'" [When Trans Persons Are Arrested to "Improve the City's Image"]. *Lifo*, July 18. http://www.lifo.gr/team/bitsandpieces/39982.
Greek National REITOX Focal Point. 2011. "Report of the ad hoc expert group of the Greek focal point on the outbreak of HIV/ AIDS in 2011." Athens: Greek Documentation and Monitoring Centre for Drugs.
Grünpeck, Joseph. 1497. *Tractatus de Pestilentiali Scorra Sive Mala de Franzos: Originem Remediaqu[ue] Eiusdem Continens*. Nuremberg: Kaspar Hochfeder.
Haider, Shuja. 2017. "The Darkness at the End of the Tunnel: Artificial Intelligence and Neoreaction," March. https://www.viewpointmag.com/2017/03/28/

the-darkness-at-the-end-of-the-tunnel-artificial-intelligence-and-neoreaction/. Accessed 31 August 2017.

Hall, Stuart, et al. 1978. *Policing the Crisis: Mugging, the State, and Law and Order*. London: Macmillan.

Hayek, Friedrich A. von. 1976. *The Mirage of Social Justice*. London: Routledge and Kegan Paul.

HCDCP (Hellenic Centre for Disease Control and Prevention). 2011. *Neōtera Epidīmiologika Dedomena gia tīn HIV Loimōxī stīn Ellada* [Latest Epidemiological Data on HIV in Greece]. http://www.keelpno.gr/Portals/0/Αρχεία/Επιδημιολογικά%20δεδομένα/HIV_ DATA_2011.pdf.

Human Rights Watch. 2013. *Unwelcome Guests: Greek Police Abuses of Migrants in Athens*. Amsterdam: Human Rights Watch.

Human Rights Watch. 2015. *Greece: Police Abusing Marginalized People*. Amsterdam: Human Rights Watch.

Ī Augī. 2016. "Dīmósia Syggnōmī Apó to KEELPNO Stis Orothetikés" [Public Apology by HCDCP to Seropositive Women], 16 November. http://www.avgi.gr/article/10813/ 7649931/demosia-syngnome-apo-to-keelpno-stis-orothetikes.

Kentikelenis, Alexander, Marina Karanikolos, Irene Papanicolas, Sanjay Basu, Martin McKee, and David Stuckler. 2011. "Health Effects of Financial Crisis: Omens of a Greek Tragedy." *Lancet* 378 (9801): 1457–58.

Khanna, Ranjana. 2003. *Dark Continents: Psychoanalysis and Colonialism*. Durham, NC: Duke University Press.

Kotouza, Dimitra. 2011. "Struggles in Greece: The Illegitimacy of Demands." *Mute* 3 (2): 115–19.

Kristeva, Julia. 1982. *Powers of Horror: An Essay on Abjection*. New York: Columbia University Press.

Kristeva, Julia. 1989. *Black Sun: Depression and Melancholia*. New York: Columbia University Press.

Lancet. 2013. "HIV Testing in Greece: Repeating Past Mistakes." *Lancet* 382 (9887): 102.

Lazzarato, Maurizio. 2012. *The Making of the Indebted Man: An Essay on the Neoliberal Condition*. Los Angeles: Semiotext(e).

Mitchell, Robert. 2016. "Biopolitics and Population Aesthetics." *South Atlantic Quarterly* 115 (2): 367–98.

Paraskevis, D., G Nikolopoulos, C. Tsiara, D. Paraskeva, A. Antoniadou, M. Lazanas, and P. Gargalianos. 2011. "HIV-1 Outbreak among Injecting Drug Users in Greece, 2011: A Preliminary Report." *Euro Surveillance* 16 (36): ii.

Pashukanis, Evgeniĭ Bronislavovich. 1980. "The General Theory of Law and Marxism." In *Selected Writings on Marxism and Law*. London: Academic Press.

Plant, Raymond. 2004. "Neo-Liberalism and the Theory of the State: From Wohlfahrtsstaat to Rechtsstaat." *The Political Quarterly* 75 (s1): 24–37.

Roberts, Dorothy. 2011. "Collateral Consequences, Genetic Surveillance, and the New Biopolitics of Race." *Howard Law Journal* 54 (3): 567–86.

Sarafis, Pavlos, Michael Igoumenidis, and Andreas Tsounis. 2013. "Exposure of HIV-Positive Sex Workers in Greece." *The Lancet Infectious Diseases* 13 (8): 649–50.

Simou, Effie, and Eleni Koutsogeorgou. 2014. "Effects of the Economic Crisis on Health and Healthcare in Greece in the Literature from 2009 to 2013: A Systematic Review." *Health Policy* 115 (2–3): 111–19.

Smith, Helena. 2014. "Greek Court Acquits Farmers Who Shot 28 Bangladeshi Strawberry Pickers." *Guardian*, July 31.

Smith, Julia. 2016. "Europe's Shifting Response to HIV/AIDS: From Human Rights to Risk Management." *Health and Human Rights Journal* 18 (2): 145–56.

Sotiropoulos, Dimitris P., John Milios, and Spyros Lapatsioras. 2013. *A Political Economy of Contemporary Capitalism and Its Crisis: Demystifying Finance*. London: Routledge.

To Vīma. 2012a. "Ī ELAS Xekiná Tis Ekkenōseis Ktiríōn Ópou Zoun Lathrometanástes" [Greek Police Begins the Evacuation of Buildings Where Illegal Immigrants Live], 25 April.

To Vīma. 2012b. "Oi Prōtoi 56 Metanástes Metaférthīkan Stīn Amygdaléza" [First 56 Immigrants Transferred to Amygdaleza], 29 April.

Triandafyllidou, Anna, and Maurizio Ambrosini. 2011. "Irregular Immigration Control in Italy and Greece: Strong Fencing and Weak Gate-Keeping Serving the Labour Market." *European Journal of Migration and Law* 13 (3): 251–73.

Tyler, Imogen. 2013. *Revolting Subjects: Social Abjection and Resistance in Neoliberal Britain*. London: Zed Books.

Weinbaum, Alys Eve. 2004. *Wayward Reproductions: Genealogies of Race and Nation in Transatlantic Modern Thought*. Durham, NC: Duke University Press.

Chapter 11

Suffocation and the Logic of Immunopolitics

Benoît Dillet

> When a transhumanist wants to change his body, like in bio-art (Stelarc), believing that the human is 'obsolete,' it is certainly possible that waste material appears in place of what is rejected—a deject (an object of rejection). But trans-jets are unconstructible. In the notion of trans-jet [*trajet*], the root *trans-* does not refer to the capacity to do whatever one wants according to whatever ends, it is the incredible apparition of a creation that shatters all prevision. *Trans-* as trans-ject . . . Contrary to the concept of subject, trans-jet implies the passage of time throughout places.
>
> —Frédéric Neyrat, *Atopies* (2014)

THE IMMUNOPOLITICS VIGNETTE

The recent election of Donald Trump, the numerous referenda in Europe (including Brexit), the problem of immigration and refugees in Europe and the new hopes for a culture of protest have challenged the status quo and urged professional political theorists to comment on, analyse and address the new political reality shaped by these earthquakes. These shocks were coproduced in part by a reorganisation of the material production of knowledge, information and news, as well as state action. Ten years after the invention of Facebook and YouTube, populist discourses have found new forms, from viral videos to push-notifications and suggested content. The Web giants have made this possible and are benefiting from these new, highly charged affective spaces. The enormous concentration of power by these new tech companies should lead us to reconsider the 'biopolitical paradigm' that is taking place alongside the restructuring of power relations is an acceleration of the politics of emotion, identity and story-telling that dominate the media to a point of saturation and suffocation. The bodies of the governed have been damaged by the economic and affective violence pouring out from a politics of avoidance and security against otherness.

In many ways, the multifaceted debate, or shall we say 'field', on biopolitics, biopower and bioethics has prefigured some of the questions that are now at the forefront of the dark futures of capitalism. As these tech companies develop new services to solve problems that were not thinkable a few years ago, they have managed to colonise life in unprecedented ways. Medicine and law are now seen as new markets for these new giants armed with algorithms, data centres, cloud services, chatbots and machine-learning gadgets.

This techno-dystopian vectorialist class disrupts management practices and produces new forms of enslavement. But what is particularly relevant for us is to examine how all this is taking place by the laying down of an 'immunopolitics', a term used by Frédéric Neyrat as an extension of Roberto Esposito's scholarly work from the early 2000s. Immunopolitics is the permanent demand for preventive measures in the search of a total immunisation from others, bacteria, terrorism, problems, or simply, negativity. But more crucially for us, beyond the logic of scapegoating and avoidance that dominates the contemporary discourse, immunopolitics also takes place at the bioeconomic, technological and ecological levels. In addition to the life instinct and the death drive central to a psychoanalytical framework, Neyrat (2011, 108) adds a third drive that he calls the 'drive to remain untouched' (*la pulsion de l'indemne*). It is the vital tendency towards an intact condition and inertia. Immunopolitics takes from this third drive by developing political strategies to operationalise it at the macrolevel. While our hyperconnected world is dominated by networks, leaks, entanglements and fluxes, a large infrastructure is being created to immunise against risks of contagion. This chapter intends to connect the literature and the question of biopolitics to that of immunopolitics. Immunopolitics rests on a dual process of a complete protection on the one hand, and absolute distancing of undesirables on the other:

> In place of bonds of community based on reciprocity, interaction and mutual expression of human relations, we would see kinds of 'auto-spheres', 'intangible bodies', via a 'distancing of the other'. This production of intangibles is based on a biopolitical split, leaving to one side untouchables, pariahs, 'disposable men' (Bertrand Ogilvie's term), the poor, immigrants, and undocumented aliens, who would be related to detention centers, prisons, or hospitals. Immunized on one side of this split, exposed on the other. (Neyrat 2011, 110)

The externalisation of other forms of life in the process of immunisation is thus crucial to the reproduction of life and the renewal of biopolitics. Transhumanism, for instance, is a doctrine particularly in fashion with

Silicon Valley gurus (or 'thought-leaders'); it claims that life can and should be enhanced to adapt to the new challenges that humanity faces. Life and technics have become one for them, by collapsing the terms of nature/culture but also nature/technics. In the introductory epigraph, Neyrat questions whether the new subjectivity that transhumanists want to put forward should not be called 'deject' instead of 'subject' given that the ultimate horizon of the techno-populists could very well be natural and human waste rather than superhumans. The question of *trans*-formation is crucial to biopolitics since it involves different understandings of life.[1]

According to Neyrat's important recent work, one of the main impetuses of the collective inertia in the face of global climate change and the drive to remain untouched is the residual humanism and productivism found in posthumanism, transhumanism and other doctrines. The protagonists of these recent intellectual movements, who write after nature and after the human, continue to argue for the self-production of humanity and use modern definitions of the human as production or *poiesis*. Beyond the logic of immunopolitics, Neyrat (2017) argues instead for an ecology of separation. At the end of my chapter, I come back to this emphasis on separation and explain why it is crucial for renewing the biopolitical and the geophilosophical paradigms. I tackle the geological unconscious of biopolitics, in showing how our conception of the living and its management by contemporary biopower are changing in our anthropocenic times.

LIFE AFTER SOVEREIGNTY?

The immense success of Michel Foucault's notions of biopolitics and biopower has had the advantage of bringing questions about life and the body, but also the inorganic, nonlife and death, to the centre of some theoretical discussions. Foucault's concept of biopolitics has led to a proliferation of neologisms such as thanatopolitics, necropolitics, neuropolitics, neuropower, psychopower, geopower and so on. By the time Foucault's work was read not only by professional political theorists but also by graduate students all around the world, the notion of biopolitics had started to be hailed as his most political and indeed politicised notion. Giorgio Agamben famously borrowed the notion for his own philosophical project of *Homo Sacer* and turned it into his own weapon by associating the biopolitical regime of modernity with the use of camps and the logic of exclusion. Agamben was struck by Foucault's nonengagement with Hannah Arendt's work and most importantly the absence in Foucault's analysis of 'the exemplary places of modern biopolitics: the concentration camp and the structure of the great totalitarian states of the twentieth century' (Agamben 1998, 4). Similarly, Achille

Mbembe (2003) has pursued Foucault's concept of biopower on other terrains in extremely productive ways. For him biopolitics is not only a regulation and management of life through the technologies of power such as birth rates, mortality rates and the development of modern statistics as part of the state apparatus. It is above all a government of death through colonial and postcolonial logics that made humans commodities and provided the framework for contemporary forms of exploitation.

Thus, the notion of biopolitics has become not only a concept but also a scheme of thought and a mode of understanding. Biopolitics is not a paradigm in the humanities and social sciences but a scheme of thought that has allowed us to overcome the abstract perspectives that hold sovereignty as central notions (Goldstone 2014). Many important studies on the biological classification of populations or on biotechnologies of reproduction used this scheme of thought. The opposition between sovereignty and biopolitics continues to dominate contemporary studies in biopolitics, which in turn demonstrates the lasting legacy of Foucault's thought given that this opposition was written in the DNA of this concept.[2] The concept of sovereignty took an imperious dimension in twentieth-century political thought and is constantly rehabilitated in the twenty-first century: 'sovereignty' appears at the same time as a problem *and* its solution, a question *and* its answer. The relative stability of this scheme of thought is surprising given the dark episodes of history from the seventeenth century onwards—I am taking Hobbes's *Leviathan* and the Treaty of Westphalia as epitomes.

Thus, some philosophers called for a renewal of the political discourse to find new conditions of possibility and impossibility. Foucault noted that it was long overdue for political theorists to cut off the king's head. But Roberto Esposito is probably the political philosopher who has voiced this concern most explicitly. Indeed, he notes that after the 9/11 attacks and the rise of security discourse (or sometimes called 'securitisation'), a new political language is required and points to Foucault's concept of biopower as this attempt to found this new thinking. To Esposito, traditional political concepts such as sovereignty, rights, democracy, representation, liberty and the individual have become inapt to understand the mutation of current politics. He writes,

> traditional political categories, such as order, but also freedom, take on meaning that forces them ever more toward the shelter of security measures. Freedom, for example, ceases to be understood as participation in the political management of the *polis* and is now recast in terms of personal security along a fault line that follows us to this very moment. (Esposito 2013, 70)

This imperative of putting 'personal security' at the forefront of all politics leads Esposito to elaborate his extremely useful couple of community/immunity, centred on the notion of the *munus* (obligation or debt in Latin).

This couple refers to two different fields: (1) the politico-legal domain organised around private property and restraints imposed on political agency; (2) the biomedical domain centred on immunology studies and practices, like inoculation. I come back to the conceptual origin of immunity in the next section, but for now it is worth remembering that this conceptual distinction is inseparable from Esposito's lucid reading of biopolitics as the knot binding politics and life together: 'Contrary to the illusions of those who imagined it was possible to retroactively skip over what for them amounted to the Nazi parenthesis so as to reconstruct the governing principles of the preceding period, life and politics are bound together in a knot that can't be undone' (Esposito 2013, 75). He notes that political philosophers all too often skip over Nazism and the redefinition of politics by Hitlerism, and pretend traditional concepts can be used as if they were transhistorical or ahistorical functions. Many theorists who analyse the life and politics knot unconsciously re-inscribe it within the frameworks of prohibition and sovereign violence. 'The problem is that sovereignty is a fiction whose basic function, by definition, has been to subsume all other fictions' (Goldstone 2014, 106). As a consequence, historical periods filled with struggles were largely overlooked. By moving from sovereignty (as a scheme of thought) to biopolitics, Esposito almost apologises for studying Nazism as a coherent political philosophy (following Emmanuel Levinas's famous and important article from 1934 [Levinas 1990]). He notes that, if we were to take seriously the relations between life and politics, we need to account for Nazism as a turning point in biopolitics. Foucault himself discussed the role of race and racist discourse in biopolitics as new political objects in his 1976 lectures *Society Must Be Defended*.[3] In this set of lectures, Foucault mainly points to the inherent racism found in socialist states. He argued that socialist states and socialist thought also make use of biopower, but that in using this new form of power, they have pursued racist politics. Much like for his 1979 lectures on neoliberalism, where he argued that the left needs to create a leftist form of governmentality (Foucault 2008, 92–94), Foucault never abhorred biopower or thought naïvely that we could step out of biopower so easily. He knew full well the advantages of the biopolitical scheme of thought, the link to the welfare state and the improvement of people's health. Hence, his last question of this lecture series is to imagine a biopolitics freed of racism.[4] This question is taken up again by Esposito in his book *Bíos* when thinking about an 'affirmative biopolitics', a concept of life freed from racism.

Numerous scholars have used the 'biopolitical paradigm' to study the state apparatus, but they often failed to historicise the use of life in politics and examine in some detail the role this concept played in slavery, colonial projects, in overtly racial societies as well as in ecological catastrophes and biotechnological innovations. Esposito is right to pose that while the constitutive/transcendental category of Marxism is history, for Nazism it is biology. But he is perhaps too quick to think that sovereignty and rights discourses can be overcome so easily by blindly following Foucault.[5]

Arguably, the biopolitical paradigm has led to radically new problems in terms of management of the living, away from older concerns of survival and obedience. This shift has taken place both at the level of governmental/state practice as well as at the theoretical level. Indeed, the discredit of Marxist philosophy of history since the 1970s and the regain of interest in vitalist strands participate in the actuality of this line of thinking. This was also partly due to the fundamental change of time and temporality in politics. We find that with the advent of the biopolitical paradigm (in theory and practice), there is a forgetting of history.

Totalising systems of thought (that used these traditional concepts that we previously listed but also make constant references to the universal and the particular, or to history and spirit) became quickly outdated and rendered inoperative to account for a real that constantly displaces itself. Political projects were being largely challenged by the vast forces of capital—but not only. The immense energies of restructuration made demands on an increasingly proletarianised population to continuously adapt to new prerogatives, cancelling or interrupting all possibilities to plan the future or to organise alternatives. In these neoliberal times, the only common denominator was therefore life: life became the centre of all contemporary politics, but given that life was an object and subject largely unaccounted for in political philosophy, a new type of political epistemology therefore needed to emerge. With the development of biotechnologies, especially reproductive medicine and human genetics, life became a new market to exploit and a new domain to regulate, legally and ethically (Waldby and Cooper 2008). Thus, from the 1970s onwards which saw the disappearance of the proletarian and class politics, a shift to life became prominent. The understanding of normality and health changed with the transformation of medicine, and the new burgeoning field of bioethics was set up to organise public debates on these issues—with limited success.[6]

BIOPOLITICS AT WORK: FROM IMMUNITY TO IMMUNOPOLITICS

This new political vocabulary—that Esposito is working with and working towards—has a longer genealogy in post-1945 continental philosophy. Indeed, Esposito's deconstruction of the political could not have been thinkable—if it is thinkable at all—without this philosophical community that worked at a vitalist philosophy against dialectical thought after 1945.[7] These new modes of thinking often attempted to account for discontinuities, fragments and events rather than long-term structures and continuities. Yet, this knot between life and politics is impossible to unravel, since when politics has come to manage and organise life, it cannot simply be undone. Esposito's new vocabulary (biopolitics, immunity, community) is crucial to grasp what is at stake the logic of immunopolitics introduced in the first section.

Concepts such as biopolitics, immunity, and community are not transcendental concepts that attempt to found a pure morality, with universal pretence, but function locally and historically. By navigating between the concepts of biopolitics, immunity, community and existential territories, we can avoid the strict oppositional thinking of the immunity/community couple. The path between the task of describing and the task of marking a world is a narrow one. Only on this path is there a possibility to add more terms: to take into consideration new factors and variables, but also to craft new political concepts.

Intuitively, we would think that if immunity is the present condition of our time that Esposito invites us to critique, community then stands as the positive (and only?) contender to replace it. Yet, things get murkier since there is a long tradition in liberal thought that opposes the individual to the community. The notion of 'community' is often problematically reduced to a liberal communitarian meaning in which sharing, in an undifferentiated way, ends up constructing a 'wider subjectivity', as Esposito (2010, 2) aptly refers to it.[8] Discourses on the community come from a nostalgic yearning for an idealised community that never was but somehow is thought to have existed. Both Jean-Luc Nancy, in his famous book *The Inoperative Community* published in 1991 (the article was originally published in 1983), and Esposito, in his 1998 work *Communitas*, begin their books by making a distinction between the liberal conception of community and a deconstructive understanding of the term.[9] Both philosophers thus expose the repetitive impulse contained in regretting the dislocation of an ideal community that never was. But this repetitive impulse is not only a foundational part of liberal thought, it is also widely shared as a constitutive principle in some sociological studies that are loosely inspired by Marxist or critical theory. Geoffroy de Lagasnerie (2012) is right that neoliberalism is often misunderstood when it is simply denounced by sociologists as a negative force (deregulation, disorder, decomposition, destruction, separation and atomisation of social life). This negative force is then often opposed to the necessity of social cohesion and collective power.

Biopolitics, much like neoliberalism, is not the dislocation of a previously tied unity but the redistribution of conceptual and practical terms in politics. It is not in itself conservative or oppressive or productive of change, it is a new operation to classify and manage biological, social and technical life. To put it differently, biopolitics and neoliberalism are often thought in terms of lack, but Foucault taught us to analyse instead the productive aspects of these new governmental rationalities. Foucault asks us to think through these rationalities before we can even think of dismantling them. While biopolitics is partly tied to new governmental practices, institutional innovations and new prerogatives (concerned with birth, mortality rates, health, disease, hygiene and so on), these practices cannot be examined independently of the breakthroughs in biology with the 'discoveries' of the genome and the DNA in 1962 and 1965, respectively. As Giuseppe Bianco and Miguel de Beistegui noted, these results and advances in molecular biology challenged the old doctrine of vitalism together with the old concept of life:

> If, after the discovery of DNA, and in the eyes of the positivist biologists such as Jacob, Monod and Lwoff, the concept of life was no longer necessary, Foucault's claim is that it did not yet exist in the eighteenth century. The life of life itself unfolds between that *no longer*, which marks the 'death' of a concept, and the *not yet*, which signals its birth. Jacob and Monod claimed that life was a useless concept in biology, Foucault was more nuanced and suggested that the historical transcendental, or the *episteme*, which dominated Western culture for 150 years—and in which a specific conception of life, labour and language played a major role—was probably destined to change. (Bianco and de Beistegui 2015, 3)

While Foucault diagnosed the end of man in *The Order of Things* in 1966, following Claude Lévi-Strauss's famous statements about the dissolution of 'man' as an operative category in anthropology, the same Foucault ten years later introduced the notions of biopolitics and biopower in relation to political economy and the state's increasing concern in people's health (Lemm and Vatter 2014). The core of the regime of sovereignty and sovereign power from the seventeenth century until the end of the eighteenth century was an 'anatomo-politics' of the governed body. With biopower, normalisation and the use of public statistics in the modern state institutions, the core moved to a biopolitics of population or species-being. The subject of politics has therefore changed from the human body to the population.

The obsolescence of the notion of life in molecular biology—replaced by the emphasis on message or code, borrowed from cybernetics—did not eradicate the notion from the other fields of inquiry; quite the contrary. In assessing the revolutions in molecular biology and the life sciences, Georges Canguilhem argued that this change of language and perspective did not take into consideration the 'subject of the knowledge of life': the biologist (François Jacob) 'ignores the *subject* who is posing the question' (Bianco and de Beistegui 2015, 6). This is where the work of the philosopher (Foucault) is crucial since it focuses on the normative power (or 'normalisation') at play in the changes of conception of life in ethics, biology as well as in politics, particularly with the creation of the welfare state. New definitions of life in life sciences have also consequences for politics and ethics; for instance, the biologist Jacques Monod 'introduce[d] the category of teleonomy and want[ed] to deduce an ethics from biology' (Bianco and de Beistegui 2015, 6). In short, while governmental agencies and political theorists have attempted to rethink the management of life institutionally but also ethically, some molecular biologists like Jacob and Monod wanted to show the political and ethical relevance of their work. Revolutions in biology made the discipline the royal science, and biological metaphors were increasingly used in social sciences and in policies to align with this scientific progress. This move was, for instance, supported by the increasing rhetoric of the 'self-organisation' of financial markets, partly by equating organisms, complex systems and impersonal rationality (Connolly 2013). However, as Catherine Malabou (2016) has recently argued, political theorists have an antibiological bias. They continue to oppose the 'natural body' to 'symbolic life', failing to see the differentiating possibilities of the living, deconstructing the understanding of the living as 'programme'. To her, even Foucault and Agamben fail to account for 'the revolutionary discoveries of molecular and cellular biology', in particular the operating categories of epigenesis and cloning, which 'are the very ones able to renew the political question' (Malabou 2016, 431).[10] This is where Esposito's concept of immunity in relation to community allows us to think the differentiating relations between life and politics.

His concept of immunity owes to Nancy's work on community. It is worth remembering that Nancy's understanding of community as *Mitsein* (being-with) draws directly from Heidegger, who had already surpassed the liberal metaphysics of subjectivity with his concept of *Dasein* (being-there). For Nancy, community is not a collection of individuals or a totality, but a being understood as a movement towards the other: a *clinamen*.[11] What defines community for Nancy is ecstasy, this experience of being outside of oneself, or in other words, the forces of this 'outside that is nothing other than the rejection of an impossible interiority' (Nancy 1991, 4). Community is not a territory and space shared in common but the experience of sharing. This experience is itself enough to weave relations and 'incompletion is its principle' (Nancy 1991, 35).

The notion of immunity, then, serves not only to associate the liberal-communitarian term of community with the desire to form a new immune system of enclaves and *ethnos*, and therefore reversing the very notion of community into immunity, but also to show that what is at stake is the entanglement of the *munus*, translated as obligation or debt. In his book *Immunitas*, Esposito traces the origin of the term 'immunity' from the Latin *immunitas* or *immunis* as what negates or lacks the *munus*, where *munus* refers to 'an office—a task, obligation, duty (also in the sense of a gift to be repaid)' (2011, 5). Thus, *communitas* is in the first instance the sharing of the *munus*, while *immunitas* is the exemption from the obligation.[12] This philological coup opens up a new field of inquiry for political philosophy. To simplify Esposito's argument to the extreme, we can say that community centres on the social life of debt, while the protection and negation of obligations define his notion of immunity. We should note that this is not primarily debt in an economic sense, but a debt in an anthropological sense, necessary to social relations. But much like Nietzsche's own commentaries on debt, this anthropological sense can be also extended to the economic sphere.

Hence, *immunitas* is the condition when someone does not repay the gift that he owes. This immunisation does not simply operate at the level of individuals as 'atoms' but also at the level of communities (in the sense of communitarians) and of society in general (institutions, generations and the arts). The challenge for Esposito is therefore to locate the processes of subjectivation at work in the *socius*. These are impersonal singularities but they are yet not independent and fully formed; the question then is: how do they hang together?

Neyrat (2010) borrows the conceptual apparatus of Esposito to think our contemporary condition as *immunopolitics*. The legal definition of immunity is the exemption to the common law: a positive norm of exception that is applicable to exceptional legal persons (politicians). Immunity is a special protection in the legal system, while in medical terms, the immunity system is what protects the body from being affected by a disease. Immunity seems at first to mark the separation between an inside and an outside but it is important to note that for both Esposito and Neyrat, immunity is what is exempted from otherness and the outside: '*immunis* is he or she who has no obligations toward the other and can therefore conserve his or her own essence intact as a subject and owner of himself or herself' (Esposito 2013, 39). Hence the new terminology that Esposito was long looking for began maturing with this community/immunity dialectic that he mobilises in his work from the 2000s. While Esposito refers to broad historical periods and traditions of thought, Neyrat on the other hand allows us to think more concretely the emergencies of the world: finance, ecology and technology.

Neyrat is the name of a formidable conceptual abstract machine—not a person but the metaphysical *topos* (*atopos*) from which endless conceptual projections gush out, an abstract machine that wants to give back the grandeur of philosophical thinking by pushing thinking to its limits. Globalisation is the project of interconnectedness that paradoxically realises the immunological drive or the drive to remain untouched. While we would think that, with globalisation, encounters with otherness are more common, for Neyrat it is nothing other than the closing of the outside: an *exophobia*. In many ways, his philosophical thought continues that of Esposito's affirmative philosophy (the impersonal, affirmative biopolitics) as well as Nancy's more recent work on politics and democracy, which I now turn to.[13] But Neyrat does this by operationalising the speculative movements of both thought to concrete problems, especially global climate change. Given the ecological emergencies, the only realist move is to develop a speculative thought. I explain why this is also connected to Neyrat's own conception of 'saturated immanence' (everything is in flux, everything is turned into flows of images, capital and other commodified 'bits').

SINGULARITIES AND THE POLITICS OF SEPARATION

What does this have to do with biopolitics? Existence means to stand outside of the self, Neyrat reminds us. Thus, the project of producing differences for the existence has consequences for biopolitics. The refusal and exclusion of all otherness and outsides (what he calls 'exophobia') from the liberal ontology means that there is no room for developing ex-centric existential territories. Neyrat suggests moving 'from a politics *over* life to a politics *of* life' (Esposito 2013, 77; Neyrat 2010, 36). Esposito's answer was to find a third term (the impersonal) beyond the person-subject ('I') as well as the infinite ethical demands of the other ('You'). Esposito's project of an affirmative biopolitics is based on life as an impersonal force, 'creating an opening to a set of forces that push it beyond its logical, and even grammatical boundaries' (Esposito 2012, 14). Rather than being a critique of the state of affairs, the impersonal is a tradition of thought, according to him, that can be recovered against a biopolitics that turns into a thanatopolitics, that is when a biopolitics defines life too strictly. Thanatopolitics and negative biopolitics begin when only a specific type of life (the proper one) can be accepted. To give content to this impersonality Esposito examines in detail three forms of the impersonal in *Third Person*: universal justice (in Simone Weil), writing (Maurice Blanchot) and life (Gilles Deleuze). Malabou would argue on the contrary that deflecting resistance to an external impersonality fails to recognise the auto-affection and self-deconstruction contained within the structure of the living (its 'plasticity'). Malabou's notion of plasticity can come close to Esposito's own conceptualisation of the immune system as 'the ever-changing product of a dynamic, competitive interaction with the environment rather than a definitive and inalterable given' (Esposito 2011, 166). It is not always clear in Esposito how his work on the impersonal fits within his immunity/community dialectic that we previously discussed.

Malabou is right to both critique the opposition between biological and symbolic life and emphasise an enriched and dynamic conception of life that, for instance, accounts for the self-repairing capacities of the body. Yet as

I noted earlier, Malabou does not see the technical life of humans, the transductive relations of the *bios* with both *technē* and *geos*. But this is also the case in Esposito's meditations on the impersonal. Indeed, what is often missing from Esposito is the question of technology, except when he turns in *Immunitas* to biotechnologies and medicine. It is precisely because the definition of immunity is widely accepted as the production of antibodies or any other immunological reaction by an organism against foreign materials or anything that is not part of that organism, that biology and politics have found a common ground. The 'dispositif of immunisation' is the edge of the spectrum that constantly threatens contemporary politics, by the promise of a sufficient amount of protection that can prevent foreign bodies from attacking. Thus, when he refers to biotechnologies, he remains at the level of metaphors to demonstrate the porosity between the biological language and the political lexicon. What he cannot think is the technicity of the human, especially in relation to nonhumans and the Earth system. Nancy, Stiegler and Gilbert Simondon have taught us that the contemporary form of life is the negotiation between nature and technics. This is a radical displacement from a humanist and modernist perspective:

> Politics is thus implicitly nothing other than the auto-management of ecotechnology, the only form of possible 'auto'-nomy that precisely no longer has recourse to any heretofore possible forms of a politics: neither the self-founding 'sovereignty', since it is no longer a matter of founding, nor the 'discussion concerning the justice' of an Aristotelian *polis*, since there is no longer a *polis*, nor even the contestation or the *differend*, since living and power go in the same direction according to an asymptomatic consensus and devoid of finality, or of truth.
>
> The term *biopolitics* in fact designates neither life (as the form of life) nor politics (as a form of coexistence). And we can certainly admit that in fact we are no longer in a position to use either of these terms in any of their ordinary senses. Both are, rather, henceforth subject to what carries them together into ecotechnology. (Nancy 2007, 94)

In this text on globalisation from 2002, Nancy presents a rare assertiveness in alerting us to the emergence of a new condition: *general equivalence*. The ills of capitalism consist in killing off the creative aspects of singularities and sense. The capitalist ontology interrupts 'world-forming' processes (*le faire-monde* in French) by reducing all things and beings to the same common denominator (exchange-value).[14] For Nancy, singularities are thus defined in the negative. For instance, enjoyment is a 'shared appropriation . . . of what cannot be accumulated or what is not equivalent, that is, of value itself (or of meaning) in the singularity of its creation' (Nancy 2007, 46). The creation of a world through the production of singularities which are not productive or accumulatable is 'the greatest risk that humanity has had to confront'

(Nancy 2007, 53), but it is also a 'task that can only be a struggle', a struggle of the West against itself. In short, for Nancy, what the discourse on biopolitics does not want to take into account is the 'becoming-technical of the world'—or ecotechnics—and this is essential since the body itself is characterised as essentially plastic and technical (Cortes Lagunas 2012). This becoming-technical of the world should be confronted by a politics of singularities as well as 'separation', as we see with Neyrat.

The power of Neyrat's work is to have integrated Nancy's concerns about the becoming-technical of the world and the critique of general equivalence as 'integral humanism' and 'saturated immanence'. The problem with biopolitics, which is in fact an ecotechnical power, is not globalisation as such but 'hominisation' as a process (Neyrat 2015a, 23). Hominisation began with the first cave paintings more than thirty thousand years ago, but even earlier than that (two to three million years ago) there were other forms of symbolisation that modified both the physical strata of the Earth as well as the sociopsychic strata of the humans as species-being (Stiegler 2017, 50). In this sense, an inquiry into ecotechnics leads us to questions about palaeoanthropology or even palaeopolitics, since the production of artefacts and technical objects is constitutive of the processes of hominisation in 'deep history'. These artefacts are themselves processes of symbolisation that constantly modify hominisation. Biopolitics and ecotechnics are therefore, for Neyrat, consequences of a broader movement that he calls 'integral humanism'—leading him to write that 'capitalism is a humanism' (Neyrat 2015a, 40–42). Philosophers who have argued against the essence of the human or anthropocentrism, whether they are humanists, posthumanists or transhumanists, tend to argue for a residual productivism. The human is nothing but can become anything or anyone, such is the slogan of the humanist tradition, revisited by the new spirit of capitalism: the entrepreneur of the self is defined by plasticity, flexibility, availability and fluidity; he or she can give himself or herself any form. The point is therefore not to argue for passivity or absolute consumption/consummation of energy and matter but to read contemporary philosophical discourses in light of the event of the Anthropocene (or Capitalocene). This is precisely the new challenge for the biopolitical paradigm, to enter into conversation with the environmental humanities and account for the fact that human life has become a geological force. Biopolitics needs to resist the *exophobia* of immunopolitics; this would consist in opposing the government of the living and the geological processes of the Earth.

This new epoch is envisaged as an event, and much like modernity is often thought as an event and a revolution in thought as well as action. In his latest book, *Defiant Earth*, Clive Hamilton (2017) argues that although environmental philosophers are writing against anthropocentrism to account for nonhuman beings and nature, they have yet to change their understanding of the environment to account for the Earth system as an ever-evolving complex totality. He uses the German term *Erlebnis* to refer to this event as a disruptive episode, recasting the German romantics and a change in feeling. In these times of acceleration, this change in feeling necessitates new regimes of symbolisation that draw from sources other than integral humanism which advocates de-inhibition (Bonneuil and Fressoz 2016) and uninjuredness (Neyrat uses the French word *indemne*). To Hamilton, what ecological thought has yet to integrate is that humans have disturbed the functioning of the Earth and this will result in immense consequences for the adaptation of humans to their new habitat and for the survival of many species. It is no longer (only) about protecting the environment and working at the microlevel but about thinking a macro-ecopolitics that constantly connects to local problems and cosmotechnics—what Félix Guattari called 'existential territories'. Guattari voiced this concern about global climate change as an event in a different vocabulary in 1989, when Earth system science was in its embryonic state:

> Throughout history and across the world existential cartographies founded on a conscious acceptance of certain 'existentializing' ruptures of meaning have sought refuge in art and religion. However, today the huge subjective void produced by the proliferating production of material and immaterial goods is becoming ever more absurd and increasingly irreparable and threatens the consistency of both individual and group existential Territories. (Guattari 2000, 46)

> Now more than ever, nature cannot be separated from culture; in order to comprehend the interactions between ecosystems, the mechanosphere and the social and individual Universes of reference, we must learn to think 'transversally'. (Guattari 2000, 43)

This transversal thinking prefigures Hamilton's call for a revolution of feeling, while his emphasis on existential territories recalls our earlier analyses of the 'creation of a world' in Nancy (a process interrupted by the general equivalence and the flat ontology of capitalism). Guattari forged the concept of 'existential territories' to think precisely of the creative processes of producing relations between singularities at the psychic and social levels. In this geological turn of political theory, in which geopower has trumped biopower as the focus, the very meaning of life that we discussed at the beginning of this chapter has also changed.

Hamilton notes that the modernist scheme of thought continues to see nature as a loving and nurturing agent (Mother Nature) instead of following Bruno Latour and other geoscientists who conceive nature as 'Gaia', a vindictive and revengeful Greek goddess:

> Our understanding of the Earth we inhabit is undergoing a radical change. The modern ideas of the Earth as the environment in which humans make their

> home, or as a knowable collection of ecosystems more or less disturbed by humans, is being replaced by the conception of an inscrutable and unpredictable entity with a violent history and volatile 'mood swings.' Earth System scientists have reached for rough metaphors to capture this new idea—images of 'the wakened giant' and 'the ornery beast', of Gaia 'fighting back' and seeking 'revenge', a world of 'angry summers' and 'death spirals'. (Hamilton 2017, 47)

Yet as Danielle Sands (2015) argues so convincingly, the problem is that all these metaphors and images reproduce gender fantasies from the Holocene: a woman who is angry, revengeful and full of 'mood swings'. This lack of imagination and creativity prevents us from grasping this 'unfolding event of colossal proportions' (Hamilton 2017, ix). To be worthy of the global climate change as an event would certainly mean to develop a vocabulary and a series of concepts that allow us to grasp the immensity of this new geological condition. The Anthropocene asks us to answer the question '"what to do" *after* "having done"' (Baranzoni 2017). Another point that is difficult to accept in Hamilton's last book is his plea for a 'new anthropocentrism' (Hamilton 2017, 50–58) that does not conform to the old humanist anthropocentrism. For him, the Earth-system thinking challenges the foundation of the dominant environmental discourse and its anti-anthropocentrism. Earth-system thinking accounts for a positive agency—the Earth as a complex system—that was missing from most strands of environmentalism.

Finally, Neyrat's ecology of separation builds on all these theories and perspectives, especially about the role of technologies and technoscientific knowledge in geopower, but without attempting to 'reset modernity' (Latour) or to bring back anthropocentrism. Neyrat's perspective is much closer to climate justice movements and the tradition of environmental justice. He begins with recognising the entanglements of beings and their increasing interconnectedness to propose a political ecology that distinguishes between forms and modes of existence. The flat ontology of capitalism poses a world without outside, without distance and without the 'capacity to draw back' (Neyrat 2017, 101):

> To be truly political, to take into consideration the dangers which may threaten us, to *distinguish* between that which humans may construct and that which *cannot* or *should not* be constructed, to know in what ways it is still possible to use the words 'nature' and 'environment,' to enable the ecosystems to be resilient and to endure the disasters of the Anthropocene, ecology must *leave space* for separation. (Neyrat 2017, 102)

However, it should be clear that an ecology of separation is not advocating a return to essentialism and modern ontological dualisms of nature/culture, nature/artifice, body/spirit, man/woman and so on. Neyrat (2016) makes the important distinction between separation and split or cleavage (*clivage*): splits do not leave room for interdependence and relations, while separations—much like the shared experience of Nancy's community—are not physical exclusions but relational disjunctions concerned with existential difference (Albernaz 2015). Neyrat's position against new materialists and eco-constructivists is not absolute but instrumental; the critique of the Great Divide was certainly useful but is not enough to think and live in Anthropocenic times.

CONCLUSION

In this chapter, I have tried to show three main things. First, I argue that the traditional opposition between sovereignty and biopower is often artificial, and that sovereignty as a scheme of thought is more pervasive than is sometimes accounted for in biopolitical studies. Second, I demonstrated how contemporary biopolitics is best thought of as immunopolitics, by following the work of Esposito and Neyrat. Modern societies and human collectives have changed their relation to the *munus*, they feel exempted from obligations and duties—and equating taxes with a burden is particularly symptomatic of this exemption. As a complement to the existing life-affirming and death drives, Neyrat noted that a third drive is central to biopolitical rationality: the drive to remain untouched. This drive is further strengthened by a large technical, political and financial infrastructure that is created by the increasing interconnectedness produced by globalisation. Nancy, on the other hand, argued that the becoming-technical of the world is the dominant feature of globalisation; the biopolitical discourse documents this becoming-technical when analysing the management of life, birth and mortality rates, reproductive medicine, post-Fordist exploitation of life as affective and digital labour and so on. Thus, Nancy poses that biopolitics should be thought in terms of ecotechnics instead—therefore, my brief turn to ecology and global climate change, since the intersectional politics to come can only take place if it reunites the four axes: race, gender, class and nature (Keucheyan 2016, 5). As Esposito's affirmative biopolitics is rather abstract and disconnected from practical politics, particularly the issues of climate change and technological disruption, it is then Neyrat's proposition for an ecology of separation that provides new theoretical insights to confront more directly the political emergencies of our time.

NOTES

I would like to thank Greg Bird and Hannah Richter for their thoughtful and perceptive comments on this chapter.

1. Bernard Stiegler notes that trance as a collective state of dreaming is fundamental to psychic and collective transformations: '*trans*-forming oneself can occur (and, even, can *only* occur) in complete unconsciousness, in assuming the unconsciousness (the improbability) necessitated by this trance [*transe*], so to speak' (2016, 132).

2. Hannah Richter discussed this aspect in the introduction in relation to the two bodies of the king in Ernst Kantorowicz.

3. For instance, Foucault defined racism as 'a break into the domain of life that is under power's control: the break between what must live and what must die' (Foucault 2003, 254).

4. Robert Bernasconi (2015) traces Georges Canguilhem's silence on the relation between science and racism in a fascinating way. Although this is probably unintentional given his well-known commitment to the French Resistance during the Second World War, his uncritical use of Kurt Goldstein and Jacob Baron von Uexküll's conceptions of the environment (as *Umwelt* and *Milieu*) makes this silence particularly problematic given their racial politics.

5. 'Biopolitics is primarily that which is not sovereignty' (Esposito 2008, 33). This is surprising, since for Foucault, the reader of Bataille (but also Esposito's reading of Bataille and Nancy), sovereignty is not reduced to the legal lexicon and can include the impersonal as well as the affirmative biopolitics that Esposito is imagining.

6. 'It is safe to say, then, that since the 1970s "life" has become a reference point for political thinking and political action in two respects. On the one hand, we can say the human "environment" is threatened by the existing social economic structures and that policy-makers need to find the right answers to the ecological question and to secure the conditions of life on Earth and the survival of humanity. On the other hand, it is becoming increasingly difficult to know, because of bioscientific discoveries and technological innovations, what exactly the "natural foundations" of life are and how these can be distinguished from "artificial" forms of life' (Lemke 2011, 27).

7. For a short overview of the deconstruction of traditional political categories in the work of Foucault, see also Judith Revel (2009).

8. The phrase is originally from the liberal communitarian political philosopher Michael Sandel.

9. 'The gravest and most painful testimony of the modern world, the one that possibly involves all other testimonies to which this epoch must answer (by virtue of some unknown decree or necessity, for we bear witness also to the exhaustion of thinking through History), is the testimony of the dissolution, the dislocation, or the conflagration of community' (Nancy 1991, 1).

10. Malabou concludes that the recent breakthroughs in molecular and cellular biology offer a power of resistance to biopolitics 'in asserting the coincidence of the symbolic and the biological. There is but one life, one life only' (2016, 438). This is beyond the scope of this chapter, but it is necessary to briefly show the limits of Malabou's own argument that verges on neurocentrism and biological reductionism. Malabou excludes *de jure* the technological question, looking for the way life repairs itself without having recourse to prostheses, grafting techniques and other incursions

of the inorganic, assuming that the body can operate a kind of self-deconstruction of its programme and identity by actualising the new biological potentials.

11. 'One cannot make a world with simple atoms. There has to be a *clinamen*. There has to be an inclination from one toward the other, of one by the other, or from one to the other. Community is at least the *clinamen* of the "individual"' (Nancy 1991, 4).

12. The community/immunity dialectic mirrors the improper/proper couple that Esposito also develops (Bird 2016).

13. Greg Bird (2016, 155) usefully writes about the main divergence between Esposito and Nancy: 'both insist that if there is a major point of contention, it is where each places their emphasis: Nancy focuses on the *cum*, while Esposito the *munus*'. Neyrat (2013, 25) uses the term 'Co-Oriented Ontology' to define Nancy's thought, as a response to the famous Object-Oriented Ontology.

14. 'Capitalism exposes the inverted form of an absolute and singular value through general equivalence' (Nancy 2007, 120).

BIBLIOGRAPHY

Agamben, Giorgio. 1998. *Homo Sacer: Sovereign Power and Bare Life*. Palo Alto, CA: Stanford University Press.

Albernaz, Joseph. 2015, November 4. "Absolved of the Absolute." *Los Angeles Review of Books*. https://lareviewofbooks.org/article/absolved-of-the-absolute/.

Baranzoni, Sara. 2017. "Anthropocenic Times: Stratigraphy of a Passage." *Azimuth* 9 (1).

Bernasconi, Robert. 2015. "The Racial Politics of Life Itself: Goldstein, Uexhüll, Canguilhem and Fanon." In *The Care of Life: Transdisciplinary Perspectives in Bioethics and Biopolitics*, edited by M. de Beistegui, G. Bianco and M. Gracieuse, 121–34. London: Rowman & Littlefield International.

Bianco, Giuseppe, with Miguel de Beistegui. 2015. "Introduction: Life: *Who* Cares?" In *The Care of Life: Transdisciplinary Perspectives in Bioethics and Biopolitics*, edited by Miguel de Beistegui, Giuseppe Bianco and Marjorie Gracieuse, 1–21. London: Rowman & Littlefield International.

Bird, Greg. 2016. *Containing Community: From Political Economy to Ontology in Agamben, Esposito, and Nancy*. Albany: State University of New York Press.

Bonneuil, Christophe, and Jean-Baptiste Fressoz. 2016. *The Shock of the Anthropocene: The Earth, History and Us*, translated by David Fernbach. London: Verso.

Connolly, William. 2013. *The Fragility of Things: Self-Organising Processes, Neoliberal Fantasies, and Democratic Activism*. Durham, NC: Duke University Press.

Cortés Lagunas, Nadia. 2012. "Sur l'écotechnie du corps." Presentation at Bernard Stiegler's Summer Academy. Épineuil-le-Fleuriel (France). https://www.youtube.com/watch?v=S6E95Q2on90.

de Lagasnerie, Geoffroy. 2012. *La Dernière leçon de Michel Foucault: Sur le néolibéralisme, la théorie et la politique*. Paris: Fayard.

Esposito, Robert. 2008. *Bios: Biopolitics and Philosophy*. Minneapolis: University of Minnesota Press.

Esposito, Roberto. 2010. *Communitas: The Origin and Destiny of Community*, translated by Timothy Campbell. Palo Alto, CA: Stanford University Press.

Esposito, Roberto. 2011. *Immunitas: The Protection and Negation of Life*, translated by Zakiya Hanafi. Cambridge: Polity.

Esposito, Roberto. 2012. *Third Person: Politics of Life and Philosophy of the Impersonal*, translated by Zakiya Hanafi. Cambridge: Polity.

Esposito, Roberto. 2013. *Terms of the Political: Community, Immunity, Biopolitics*, translated by Rhiannon Noel Welch. New York: Fordham University Press.

Foucault, Michel. 2003. *"Society Must Be Defended": Lectures at the Collège de France, 1975–76*, translated David Macey. New York: Picador.

Foucault, Michel. 2008. *The Birth of Biopolitics: Lectures at the Collège de France 1978–79*. Basingstoke, UK: Palgrave Macmillan.

Goldstone, Brian. 2014. "Life after Sovereignty." *History of the Present* 4 (1): 97–113.

Guattari, Félix. 2000. *The Three Ecologies*, translated by Ian Pindar and Paul Sutton. London: The Athlone Press.

Hamilton, Clive. 2017. *Defiant Earth: The Fate of Humans in the Anthropocene*. Cambridge: Polity Press.

Keucheyan, Razmig. 2016. *Nature Is a Battlefield: For a Political Ecology*, translated by D. Broder. Cambridge: Polity Press.

Lemke, Thomas. 2011. *Biopolitics: An Advanced Introduction*, translated by Eric Frederick Trump. New York: New York University Press.

Lemm, Vanessa, and Miguel Vatter, eds. 2014. *The Government of Life: Foucault, Biopolitics and Neoliberalism*. New York: Fordham University Press.

Levinas, Emmanuel. 1990. "Reflections on the Philosophy of Hitlerism," translated by Séan Hand. *Critical Inquiry* 17 (1): 62–71.

Malabou, Catherine. 2016. "One Life Only: Biological Resistance, Political Resistance," translated by Carolyn Shread. *Critical Inquiry* 42 (3): 429–38.

Mbembe, Achille. 2003. "Necropolitics." *Public Culture* 15 (1): 11–40.

Nancy, Jean-Luc. 1991. *The Inoperative Community*, translated by Peter Connor, Lisa Garbus, Michael Holland and Simona Sawhney. Minneapolis: University of Minnesota Press.

Nancy, Jean-Luc. 2007. *The Creation of the World, or Globalization*, translated by François Raffoul and David Pettigrew. Albany: State University of New York Press.

Neyrat, Frédéric. 2010. "The Birth of Immunopolitics," translated by Arne de Boever. *Parrhesia* 10: 31–38.

Neyrat, Frédéric. 2011. "Intact," translated by Roxanne Lapidus. *SubStance* 40 (3): 105–14.

Neyrat, Frédéric. 2013. *Le Communisme existentiel de Jean-Luc Nancy*. Paris: Nouvelles Éditions Lignes.

Neyrat, Frédéric. 2014. *Atopies: Manifeste pour la philosophie*. Caen: Éditions Nous.

Neyrat, Frédéric. 2015. *Homo Labyrinthus: Humanisme, Antihumanisme, Posthumanisme*. Arles: Éditions Dehors.

Neyrat, Frédéric. 2016. 'The Biopolitics of Catastrophe, or How to Avert the Past and Regulate the Future'. *South Atlantic Quarterly* 115 (2): 247–65.

Neyrat, Frédéric. 2017. “Elements for an Ecology of Separation: Beyond Ecological Constructivism.” In *General Ecology: The New Ecological Paradigm*, edited by Erich Hörl with James Burton, 101–28. London: Bloomsbury.

Revel, Judith. 2009. “Identity, Nature, Life: Three Biopolitical Deconstructions.” *Theory, Culture and Society* 26 (6): 45–54.

Sands, Danielle. 2015. “Gaia, Gender, and Sovereignty in the Anthropocene.” *philoSOPHIA* 5 (2): 287–307.

Stiegler, Bernard. 2016. *Automatic Society*, vol. 1: *The Future of Work*, translated by Daniel Ross. Cambridge: Polity Press.

Stiegler, Bernard. 2017. *Philosophising by Accident: Interviews with Élie During*, translated by Benoît Dillet. Edinburgh: Edinburgh University Press.

Waldby, Catherine, and Melinda Cooper. 2008. “The Biopolitics of Reproduction: Post-Fordist Biotechnology and Women’s Clinical Labour.” *Australian Feminist Studies* 23 (55): 57–73.

Index

About the Contributors

Hannah Richter is Lecturer in Politics and International Relations at the University of Hertfordshire, UK, as well as PhD candidate in Political and Social Thought at the University of Kent, UK. In addition to her dissertation, which develops a political theory of sense drawn from the philosophy of Gilles Deleuze and Niklas Luhmann's sociological systems theory, she is working on biopolitics, relational ontologies and Deleuzian conceptions of time, event and resistance. Her analysis of immunity in Roberto Esposito and Niklas Luhmann has been published in the *European Journal of Political Theory*.

Marco Piasentier is Postdoctoral Researcher at the University of Helsinki, and Research Fellow at the Centre for Critical Thought, University of Kent. He is the coeditor of a special issue of *Paragraph* entitled 'Italian Biopolitical Theory and Beyond: Genealogy, Psychoanalysis, Biology', and the author of various articles on modern and contemporary continental philosophy. He is completing a monograph entitled *Between Language and Life: The Question of Biopolitics*. Marco is also part of the editorial board of *Occam's Razor: Philosophy/philosophies* (Editorial Group L'Espresso) and of the *Journal of Italian Philosophy*. Since 2015 he is the codirector (with Andrea Soardo) of the International School of Philosophy (ISP/Grado).

Jemima Repo is Lecturer in the Politics of Gender at Newcastle University, UK, and a Visiting Researcher at the University of Helsinki working on the Academy of Finland–funded project 'Biopolitics and Democracy'. Her research interests include feminist and gender theory, biopolitics, the politics of population, human capital theory and the politics of popular culture. Her work has appeared in journals such as *Social Politics*, *European Journal of*

Cultural Studies, *Feminist Theory*, *International Feminist Journal of Politics*, *Theory & Event* and *European Journal of Women's Studies*. Her book *The Biopolitics of Gender* (2015) was awarded the 2017 International Studies Association Feminist Theory and Gender Studies Book Prize.

Hidefumi Nishiyama is Postdoctoral Fellow at the RELATE Centre of Excellence, Academy of Finland, and in the Department of Geography at the University of Oulu, Finland. He received his PhD in politics and international studies from the University of Warwick, UK, in 2016. He is currently working on a book project on the genealogy of racism, biopolitics and biometrics in modern Japan, of which his chapter in this volume forms a part.

Greg Bird is Assistant Professor of Sociology at Wilfrid Laurier University, Canada. His research unpacks the philosophical underpinnings and the practical implications of immunitarian biopolitics, particularly with regard to the governance of migration. He has extensively published on Italian theory, in particular the work of Roberto Esposito, such as the monograph *Containing Community: From Political Economy to Ontology in Agamben, Esposito and Nancy* (2016) and *Community, Immunity and the Proper: Roberto Esposito* (2015, co-edited with Jon Short). Greg Bird is cofounder and coordinator of the interdisciplinary research group *Techne*, focused on advancing biopolitical research at Wilfrid Laurier University, and a member of the international research network 'Workiteph: Network on Italian Thought and European Philosophies'.

Maria Fannin is Reader in Human Geography in the School of Geographical Sciences at the University of Bristol, UK. Her research focuses on the social and economic dimensions of health, medicine and technology, particularly in relation to reproduction and women's health. She has conducted research on commercial cord blood banking, conceptualisations of hoarding and exchange in the biological tissue economy, and feminist geographical approaches to a 'bodily commons' in a postgenomic age. She is currently researching the multiple forms of value attributed to human placental tissue in the biosciences, medicine and alternative health practices. Her work has appeared in *Body & Society*, *Feminist Theory*, *New Genetics & Society* and *Gender, Place and Culture*.

Charlie Yi Zhang is Assistant Professor in the Department of Gender and Women's Studies at the University of Kentucky. His research brings feminist, queer, critical race and affect studies scholarship to the forefront of the debate over neoliberal globalization. His publications focus on the topics of nationalism/globalism and hypermasculinity, gendered and racialized neoliberal restructuring in China and the Asia-Pacific region, the queering resistance to neoliberal governance, and critical theory of body and the US electoral politics. Tentatively titled *Crossing Waters between National and Global—The Lovescape that Enables China*, his current book project offers to develop a theoretical and methodological framework of love to shed light on the mechanism that has fuelled neoliberal globalization and its recent hate-filled nationalist/antiglobalist backlashes, and explore possibilities of grounded and substantive social changes.

Christian Klesse is Reader at the Department of Sociology at Manchester Metropolitan University. Christian studied social and economic history, anthropology, politics, contemporary German literature and gender and ethnicity studies in Germany and the UK and holds a PhD in sociology from the University of Essex. His research interests lie in the fields of gender and sexual politics, including nonmonogamy and polyamory, transnational LGBTQ activism and queer film festivals. His work is interdisciplinary and often collaborative. He is author of *The Spectre of Promiscuity: Gay Male and Bisexual Non-monogamies and Polyamories* (2007). Recent journal articles include 'The Politics of Age and Generation at the GAZE International LGBT Film Festival in Dublin', with Jon Binnie, *The Sociological Review* (2017); 'Theorising Multi-partner Relationships and Sexualities—Recent Work on Non-monogamy and Polyamory', *Sexualities* (2017); and 'Marriage, Law and Polyamory. Rebutting Mononormativity with Sexual Orientation Discourse?' *Oñati Socio-Legal Series* (2016).

Shubranshu Mishra is Lecturer in Politics and International Relations at the University of Exeter. He received a PhD from the University of Kent for his dissertation titled *Bearing Witness: Truth, Violence and Biopolitics of Everyday Lives in Kashmir*, which explored the heterogeneities of bearing witness to bring to light the continuum of impunity in Kashmir and structures of militarisation and acquiescence. His research interests lay at the intersection of decolonial thought and biopolitical projects to highlight the colonial underpinnings of world politics and international relations. Previously, he has worked with Amnesty International in India and holds postgraduate degrees from Jawaharlal Nehru University and Jamia Millia Islamia, New Delhi.

Dimitra Kotouza is Early Career Researcher and currently Lecturer in Sociology at Middlesex University, UK. She received her PhD in Political and Social Thought from the University of Kent and is the author of *Surplus Citizens* (forthcoming). Her work questions dominant Eurocentric and ethnocentric understandings of the economic and social crisis in Greece, and examines class, racialised and gendered entanglements of collective practice in this conjuncture. She is also a coeditor of *Mute*, a journal on art, politics and technology.

Benoît Dillet is Teaching Fellow in French Politics at the University of Bath, UK. He is the coauthor of *The Political Space of Art: The Dardenne Brothers, Arundhati Roy, Ai Weiwei and Burial* (2016), coeditor of *The Edinburgh Companion to Poststructuralism* (2013) and *Technologiques: La Pharmacie de Bernard Stiegler* (2013), and translator of Bernard Stiegler's *Philosophising by Accident: Interviews with Élie During* (2017). He is also one of the editors of the online journal *La Deleuziana*.

www.ingramcontent.com/pod-product-compliance
Ingram Content Group UK Ltd.
Pitfield, Milton Keynes, MK11 3LW, UK
UKHW041103150926
475561UK00001BA/24

9 781786 602701